Practical Interface Circuits for Micros

George Loveday
CEng, MIERE
Senior Lecturer in Electronic Engineering
Bromley College of Technology

Pitman

PITMAN PUBLISHING LIMITED
128 Long Acre, London WC2E 9AN

Associated Companies
Pitman Publishing New Zealand Ltd, Wellington
Pitman Publishing Pty Ltd, Melbourne

© G C Loveday 1984

First published in Great Britain 1984

All rights reserved. No part of this publication may be reproduced, stored in a retrieval system, or transmitted, in any form or by any means, electronic, mechanical, photocopying, recording and/or otherwise without the prior written permission of the publishers. This book may not be lent, resold, hired out or otherwise disposed of by way of trade in any form of binding or cover other than that in which it is published, without the prior consent of the publishers. The book is sold subject to the Standard Conditions of Sale of Net Books and may not be resold in the UK below the net price.

ISBN 0 273 01998 8

Text set in 10/12 pt Linotron Times,
printed in Great Britain
at The Pitman Press, Bath

Contents

Preface

1 Systems 1
1.1 The systems model 1
1.2 The interface problem 2
1.3 Types of system 10
1.4 An interfacing example – oven temperature control 14

2 Useful Electronic Devices and Circuits 26
2.1 Electronic component ratings 26
2.2 Semiconductors – diodes, transistors, FETs 27
2.3 Linear/analog integrated circuits 36
2.4 Power-switching devices 40
2.5 Useful analog circuits 45

3 Digital Circuits 52
3.1 Introduction 52
3.2 Logic gates 53
3.3 Boolean algebra 55
3.4 Logic conventions and parameters 58
3.5 Logic families 59
3.6 Interfacing between logic 63
3.7 Tri-state logic 64
3.8 Bistables 65
3.9 Shift registers 69
3.10 Counters 70

4 Sensors and Output Devices 74
4.1 Transducers – an introduction 74
4.2 Temperature sensors 76
4.3 Light sensors and devices 83
4.4 Position and force sensors 87
4.5 Ultrasonic devices 90
4.6 Output devices 92

5 Conversion 103
5.1 Data conversion 103
5.2 Digital-to-analog conversion 105
5.3 Practical DACs 108
5.4 Analog-to-digital conversion 116
5.5 Practical ADCs 117
5.6 Sample-and-hold circuits 126

6 Microprocessors and Interface Adaptors 127
6.1 Types of interface adaptor 127
6.2 The 6800 range of microprocessors 128
6.3 Architecture and programming the 6800/6802 130
6.4 Program example for the 6800/6802 139
6.5 The PIA type 6821 140
6.6 A simple PIA application 147
6.7 The asynchronous communications interface adaptor M6850 151

7 Application Examples 158
7.1 Introduction 158
7.2 Using switches 158
7.3 Driving displays 166
7.4 A general-purpose DAC board 170
7.5 A general-purpose ADC board 171
7.6 System example 174

8 Interface Exercises 184
8.1 An 8 by 8 LED matrix display 184
8.2 Waveform generation using the DAC board 184
8.3 D.C. motor control reversal 186
8.4 D.C. motor control – addition of counter 186
8.5 Light level control 187

Index 188

Preface

The art of interfacing a microcomputer or a microprocessor within a control system requires an understanding of both software and hardware techniques. In this book I have attempted to integrate both of these aspects to give an introduction to the more practical methods involved in interfacing. The text includes chapters on: electronic devices and circuits used in interfacing; sensors and output devices; data convertors (A to D and D to A); interface adaptors and microprocessors; and system examples. My aim has been to make the text suitable for programmers who have a wish to extend their knowledge of electronic components and systems, and for electronic engineers and students who may need to broaden their knowledge of microprocessors and software techniques. In view of this there may be topics which will effectively be revision for some readers.

Further information on electronic devices, components and techniques can be found in *Essential Electronics* (Loveday) and on microprocessors in *Microprocessors: Essentials, Components, Systems* (Meadows and Parsons), both published by Pitman.

Throughout the text the following symbols have been used:
    # for immediate mode
    $ for hexadecimal numbers
I have also taken to spelling Analogue as Analog.

G. L. June 1984

1 Systems

1.1 The Systems Model

Suppose we need to connect up a microprocessor or microcomputer as the controlling element of some process; perhaps several different devices such as input sensors, amplifiers, convertors and output drivers will be required. Having assembled the various elements and made all the correct connections we will then have a *system*, that is to say a collection of various component parts all working together as a whole. The problems encountered in setting up the system and in getting it running will be mainly concerned with

1 Achieving the correct interconnection method between the various parts.
2 Ensuring that signals from any one part of the system are compatible with those parts to which it is connected.
3 That a proper set of instructions with correctly arranged timing are programmed into the microcomputer.

The first two points are about the types of interface circuitry to be used, that is the circuits which link the various elements, while the third depends upon the quality of the software, i.e. the program.

A good first approach to any problem such as this, *before* beginning to write the program, is to follow a logical design sequence:

a) Define the exact task that the system has to carry out.
b) Sketch a block diagram of the essential parts of the system.
c) Produce a flowchart of the actions required by the system.

Take the example of an oven controller in which the oven temperature has to be maintained at a preset value set by a reference input. For a start we shall assume that only a simple type of control is required, but obviously a micro system would be capable of very sophisticated and subtle control of the heating process.

The task as already defined is to maintain the temperature in the oven as close as possible to the desired value, and *fig. 1.1* shows the basic block diagram of the proposed system. The temperature inside the oven has to be sensed by some kind of transducer (a device for converting one form of energy into another). In this case, either a thermocouple or a thermistor with bridge arrangement would be suitable to convert the heat energy into an electrical signal. An amplifier is necessary to increase the size of the relatively weak transducer signal; and before being sampled by the microcomputer this varying signal (an analog signal) must be converted into a suitable digital form. The digital word applied to the microcomputer will then be proportional to the temperature level in the oven.

2 Practical Interface Circuits for Micros

Fig. 1.1 Block diagram of a heater system

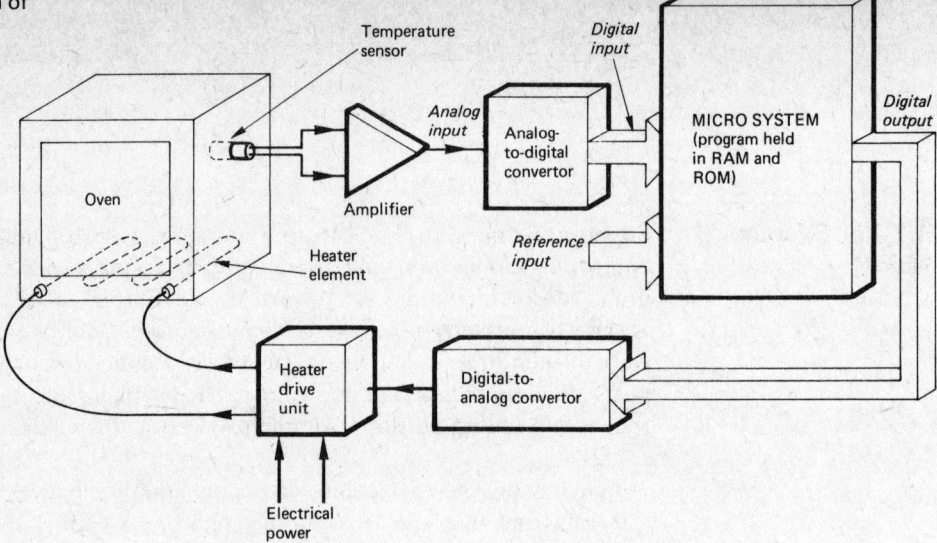

The instructions contained in the program memory of the microcomputer will cause this digital signal to be sampled and compared with the reference input (a digital word set to be proportional to the desired temperature). Depending on the result of this comparison, the microcomputer will output a digital command to adjust the amount of electric power applied to the heater. The flowchart for the various steps required in the program is shown in *fig. 1.2*.

What must be noted from this initial outline of our temperature control system is that, although providing an excellent starting point for a design, it in no way gives enough detail to allow the systems to be fully implemented. Also it only shows one of several methods of carrying out the task. The type of sensor, amplifier, analog-to-digital convertor and heater drive unit are not closely specified. In each case there are several different devices, circuits and techniques to choose from. It is at this stage that the most difficulty in getting the system operational may be experienced. Some skill and knowledge is required to get the various pieces of hardware to operate at the optimum level. The following sections deal with the problems which arise in this design area, that is in the interface circuits and input/output devices.

1.2 The Interface Problem

What is the definition of the word interface? It could be taken as

a) "The boundary between two regions", as for example the beach is the interface between the land and the sea, or

b) "The area in which the interactions take place between any two connected parts in a system".

Fig. 1.2 Flowchart for heater system

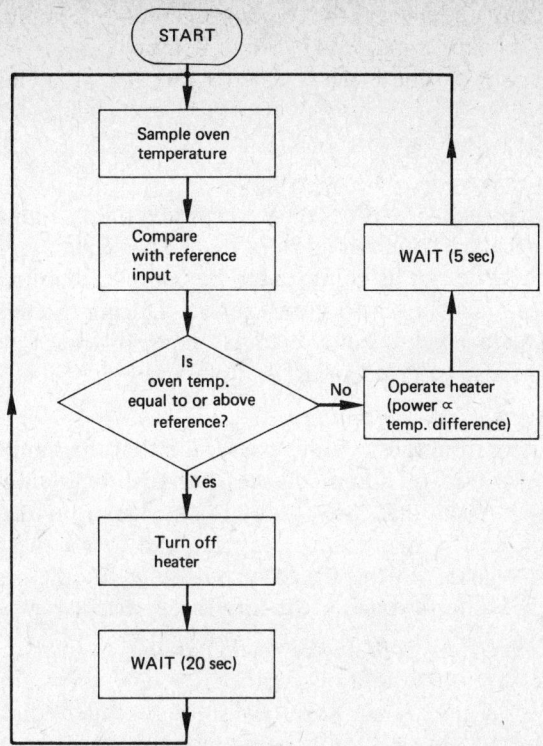

For our purposes the latter is more appropriate and an interface can be taken to mean any electronic circuit that takes a signal from one part of a system and adjusts it so that the resulting signal is fully compatible with another part of the system. This is illustrated in *fig. 1.3* where the interface receives a varying signal from device A and converts it into a suitable form (in this case digital) to match the input requirements of unit B.

It can be seen, then, that a wide variety of circuits fall within the range of the description "interface" and that in fact a particular interface circuit could well be unique to one system. This extensive range of circuits and devices to choose from could be said to be the *interface problem*. The solution to this ever-expanding range of unique circuits and devices is to study techniques, for very often an interface called (x) uses the same technique in achieving a signal match or translation as a circuit (y). Although the circuits and connections may appear entirely different, the technique, i.e. the method used to get the match, is the same.

Fig. 1.3 Basic interface

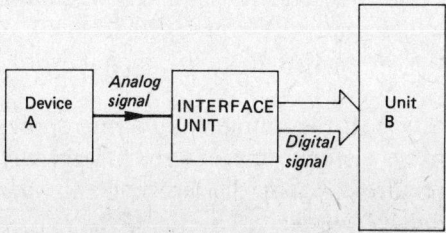

Having understood the various techniques it is simply a matter of adapting them to suit a particular need. Some of the commonly used interfacing techniques can be broadly listed under the following headings. A few brief explanatory notes are provided here but each technique will be covered in much more detail further on.

1 *Multiplexing and sampling*

Multiplexing is the process by which several data inputs can be switched, one at a time, to a common line (*fig. 1.4*). The switch, an electronic type, operates at a set speed or as commanded by computer control and therefore samples each input for a defined time period. During the time between samples, the input data level is held relatively constant ready for conversion. This last point is also referred to as "sample and hold".

2 *Sequencing and timing*

It follows from the previous paragraph that the sampling of input sensors and the operation of output devices has to be carefully controlled so that the required tasks are carried out in the arranged order and in correct time sequence. The microprocessor itself can be set the task of generating these timing signals via the program but quite often, to save computer time, an external timer as part of the interface is used.

3 *Conversion from analog to digital and vice versa*

To be acceptable to a microprocessor, any varying input signal must be first converted into some form of suitably coded digital word. Suppose the temperature in an oven is being monitored using a temperature transducer. After amplification, the voltage level proportional to temperature is applied to an analog-to-digital convertor (ADC), as shown in *fig. 1.5*. If a 4-bit digital word is used, a table of conversions could be drawn up as follows:

Temperature (°C)	Analog input to ADC (V)	Digital output from ADC MSB LSB
20	1	0 0 0 1
40	2	0 0 1 0
60	3	0 0 1 1
80	4	0 1 0 0
100	5	0 1 0 1
300	15	1 1 1 1

Here, because only 4 bits exist in the digital output, each 20°C increase in temperature changes the digital output by 1 (one). The varying analog input has been split into 16 discrete levels and it would not be possible using only 4 bits to resolve temperatures closer together than 20°C. In any conversion, the greater number of bits used in the digital word, the higher will be the resolution.

In the same way, if the output from a microprocessor is required to drive some analog device, for instance a chart recorder, then a digital-to-analog convertor is required. Again, the larger the number of bits, the finer will be the resulting analog output.

Fig. 1.4 Multiplexing

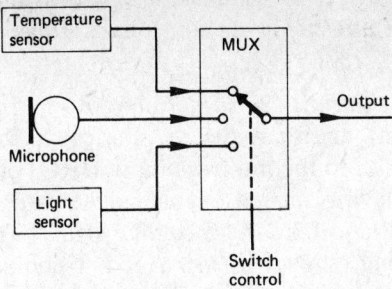

Fig. 1.5 Principle of analog-to-digital conversion

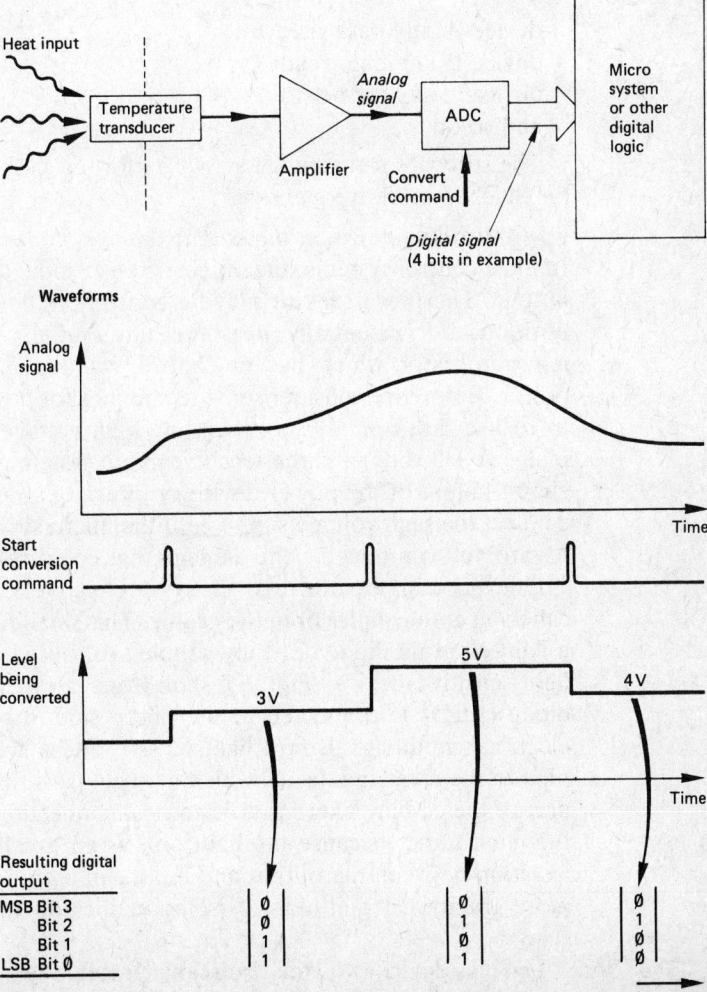

Well-established techniques exist for the task of converting from analog to digital and vice versa, and several ICs are available, most of which are provided with facilities allowing control via a microprocessor.

4 *The use of interrupts or polling*

An **interrupt** is a useful way of getting the microprocessor to service (i.e. accept data from) an input device. Basically, the device sends a signal requesting attention to the interrupt input (IRQ) of the microprocessor. The microprocessor having carried out its current program instruction will acknowledge the call and take the incoming signal. This is the basis of what is called **handshaking**, shown in *fig. 1.6*, a technique that is also used for outputting from the microprocessor to a device.

When there are several input devices to the system, an alternative method called **polling** can be used. In this method the microprocessor, via the program, checks each device in turn until it finds a device which is requesting attention. That is,

device A any data ready?
device B any data ready?
device C any data ready?
and so on

The software can arrange a *priority* for the polling so that fast devices can be checked more frequently.

5 *Controlling power at the output*

In most control systems, the object is to get some actual work achieved at the output. The power device may be controlling flow, heat, light intensity, or position, but essentially it is the tiny output signal available from the microprocessor which has to control, via the interface, the large output power. Before we consider some techniques for power control, it is important to realise that *isolation* between any high voltage supply (typically the a.c. mains at 240 volts) and the sensitive micro system is a vital feature. Any short circuit failure of the power device or interface circuit must not be allowed to connect the high voltage supply into the micro system. Apart from the safety hazard you can imagine the damage that could result.

One popular method for achieving effective isolation is to use a device called an **opto-coupler** or **opto-isolator**. This small fully-enclosed IC consists of a light-emitting diode optically coupled to, but electrically insulated from, a light-sensitive device. *Fig. 1.7* shows one arrangement. The light-emitting diode (LED) is connected to the micro side of the interface and converts electrical input signals into light pulses. These are transmitted via the light pipe to the receiving device which then controls the input to the high voltage part of the circuit. There is no electrical connection between the two parts of the circuit and, because the light pipe is an excellent insulator, the voltage isolation between the output and input can be in excess of 4 kV. Alternative more traditional methods of isolation include the use of transformers or relays.

Having dealt with the isolation problem we can now look at some techniques for power control.

Systems 7

Fig. 1.6 Use of interrupts

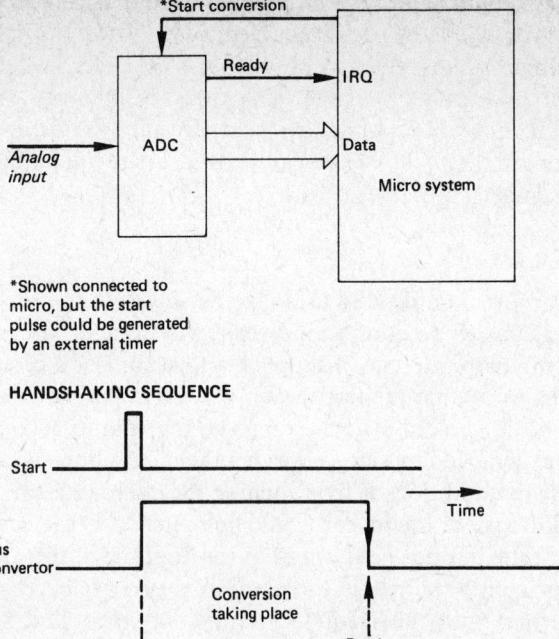

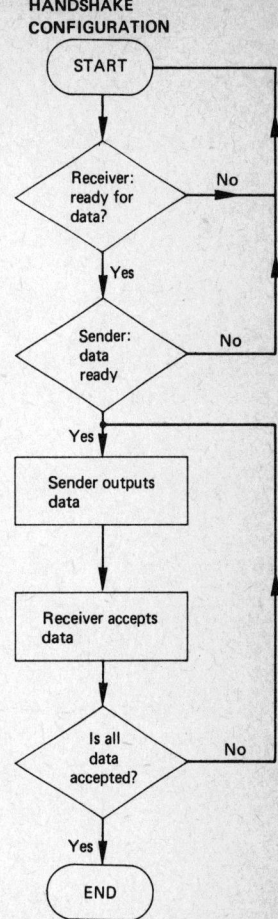

Fig. 1.7 Use of opto-coupler for isolation

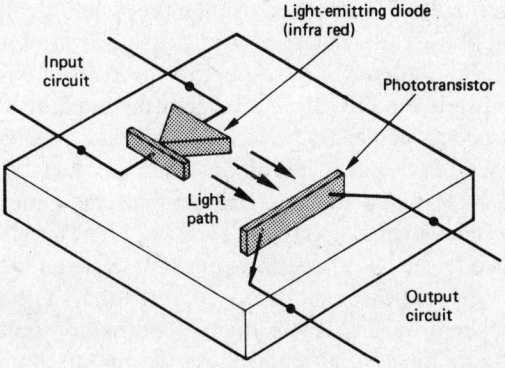

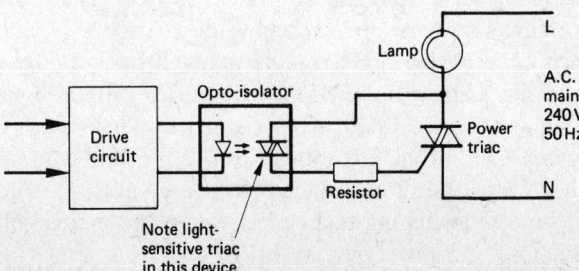

1 Control in an ON/OFF switching situation is relatively straightforward. The interface will consist of an electronic power switching device such as a Darlington transistor or VMOS power FET. This will be driven fully ON when the micro outputs a logic 1 signal and it will be OFF when the micro output is logic $\emptyset$. The main points are to ensure that the device can pass the required current in the ON state and can withstand the full voltage in the OFF state. This is discussed fully later.

2 A more complicated arrangement has to be set up when the power in the load needs to be controlled over a wide range, and the techniques involved depend upon the type of supply being used. For d.c. situations it is best to use a pulse width modulation system (PWM). Suppose the speed of a d.c. motor is to be controlled over the range 60 rev/min to 3000 rev/min (*fig. 1.8*). This can be achieved by simply varying the d.c. voltage applied to the motor; but a PWM method does it by switching the motor on for a variable time period. For low speed, the motor would be switched to the d.c. supply for a relatively short time in one cycle, whereas for high speed the motor would be switched on for nearly the whole time period in one cycle. The width of the switching waveform and the frequency could be controlled via the program in the microprocessor.

3 A.C. power control is achieved using either what is called phase control or burst firing. In **phase control**, a trigger pulse is generated with its time position variable (set by the micro) with respect to the start of each mains half-cycle. This trigger pulse is used to "fire" an electronic latching switch such as a triac which then connects power to the load. As the mains waveform goes through zero, the latching device switches off and then conducts again at a point in the negative half-cycle when the next trigger pulse is generated. This is more easily seen from a waveform diagram as in *fig. 1.9*. This shows that the greater the delay between the start of each mains half cycle and the trigger pulse, the smaller will be the power supplied to the load. The delay time can be controlled at the interface by a digital command signal from the microprocessor. The method is particularly useful in lamp dimmers and a.c. motor speed control.

Where a load has a much slower response, heaters being an obvious example, the **burst firing** technique can be used. In this method, the load is connected to the mains supply by the switching device for a set number of whole cycles. (Each cycle in the 50 Hz mains takes 20 msec.) Thus for low power dissipation in the load, the switch may be on for only a few cycles in every hundred (10 in a 100 gives 10%), whereas fifty cycles in every hundred would give half power. One of the advantages of this method compared with phase control is that very little interference (electrical "noise") is generated because the actual time of switching takes place when the mains voltage has just passed through zero. This is shown in *fig. 1.10*.

Systems 9

Fig. 1.8 Principle of pulse width modulation control

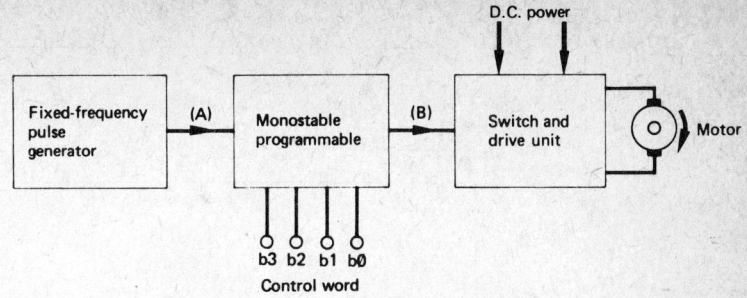

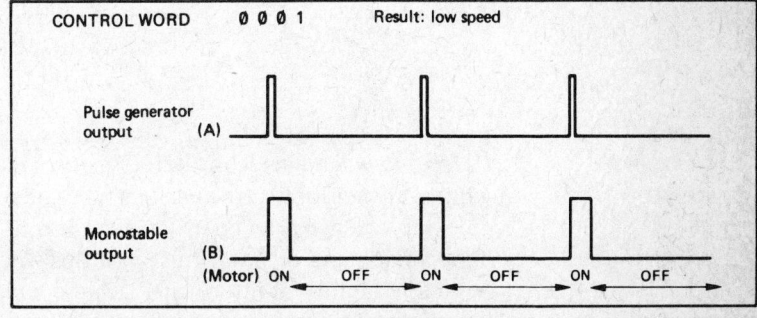

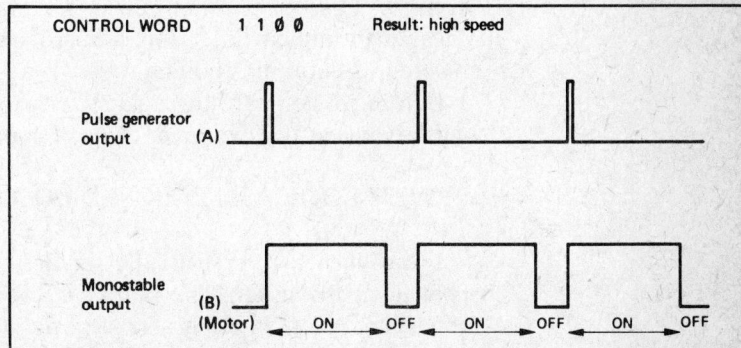

Fig. 1.9 Phase control in a.c. power circuits

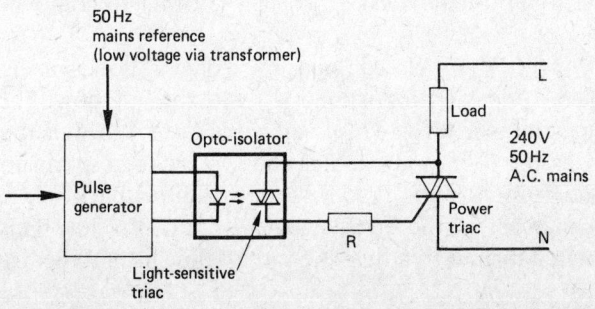

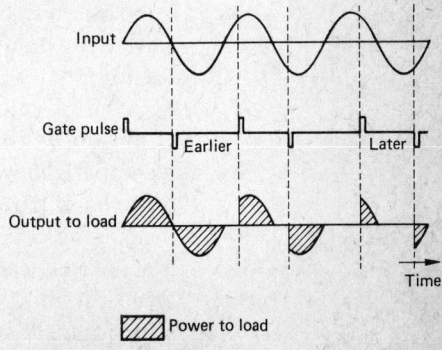

Fig. 1.10 Burst firing

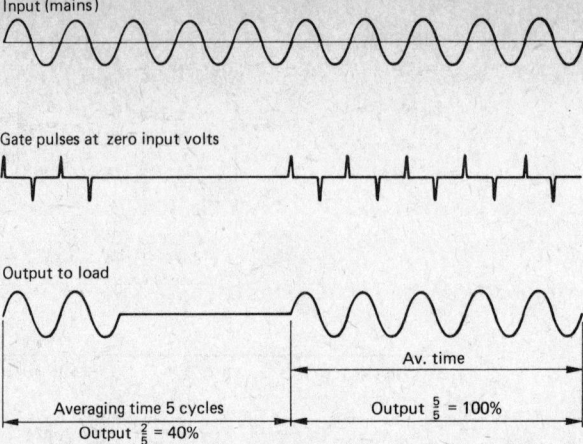

1.3 Types of System

Up to now we have considered systems mainly from the digital point of view, but a further look at the various types of system and the way in which they can be implemented can be instructive. It is useful to make a comparison between the analog and digital forms since this will assist in making the decision about the advisability of interfacing a microcomputer or microprocessor in the first instance. In some situations an analog system may in fact prove simpler and cheaper to implement. Using a micro system for some trivial task could be wasting its enormous potential.

Before we compare the analog versus digital approaches, we must distinguish between the two main types of control. These are

OPEN LOOP   and   CLOSED LOOP

In an **open loop system**, the required output is set by a reference level applied at the input. The output is completely unaffected by the result it produces. An example is shown in *fig. 1.11a* where the input voltage level from the potentiometer sets, via the controller, the output valve to a desired position. It is assumed that, with the valve set to this position, a certain fixed amount of liquid flow will take place; but this will only be the case if conditions at the output do not vary. Should, for any reason, the viscosity of the liquid in the pipe increase, the rate of flow would fall, and the open loop system would not be able to adjust the valve's position in order to compensate.

The system can be converted to **closed loop** by using a flow-sensing transducer that feeds back a signal proportional to the rate of flow, for comparison with the input reference level. This is shown in *fig. 1.11b*. It is the **feedback** path which closes the loop and enables the system to respond to changing conditions at the output. If the flow rate falls, the feedback signal decreases, causing a net increase in the error signal to the controller. This causes the valve to be opened further to adjust the rate of flow back to nearly its original value.

Fig. 1.11a Open loop system

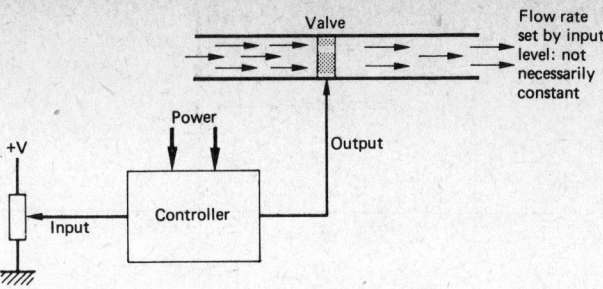

Fig. 1.11b Closed loop system

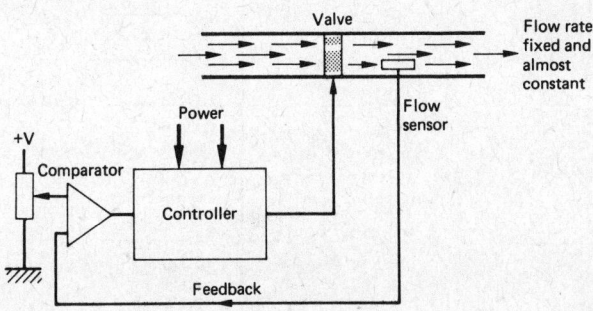

A closed loop system is inherently more accurate than an open loop type but is more complex and has more of a tendency to instability. This instability can result if too much gain is set into the error amplifier. Fast response would be achieved but the valve's position might move backwards and forwards for a few cycles before settling down following an input of a new desired position. Of course this would depend upon the inertia of the output arrangement and there are relatively simple techniques for ensuring that a closed loop system remains stable.

The system block diagrams of *fig. 1.11* are completely analog (linear) but both the open loop and the closed loop forms can also be created using microprocessor digital controllers. What then are the differences between analog and digital systems? We start by looking at signals. A true **analog signal** is one that is of a continuous nature; in other words, it is a quantity that can take any value within defined limits. The majority of inputs to systems—the electrical signals from transducers—are therefore analog. This is because most of the energy forms which have to be measured for control purposes, such as heat, force, velocity, pressure, and light intensity, are all continuously variable and not discontinuous. Take the output from a microphone; it is an electrical voltage which follows (i.e. is analogous to) the input sound wave (*fig. 1.12*).

In contrast, a **digital signal** is discontinuous and consists of either a coded serial pattern or a parallel group of discrete voltage levels. Commonly, only two levels are used:

> On/Off   Low/High   Logic $\emptyset$/Logic 1

Fig. 1.12 The difference between analog and digital signals

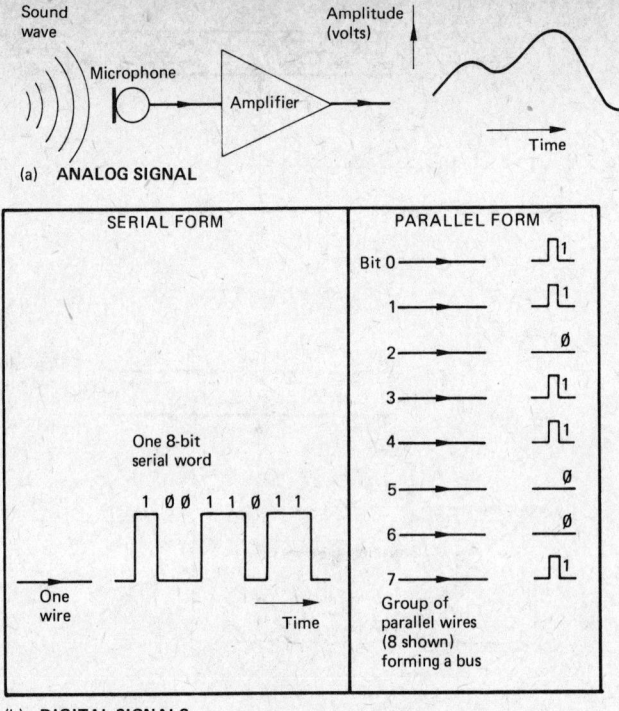

Figure 1.12 illustrates the difference between the two kinds of signal, showing how a small time section of an analog waveform could be digitised. Obviously, the conversion requires a certain amount of hardware (ADC in this case), making the digital system more complex than the analog, and some information detail contained in the analog signal will be lost in the conversion process. This loss of resolution can be minimised by using more bits in the digital word, but a finite loss always exists. To counteract these points, the digital system has several distinct advantages over the purely analog form:

a) The digital signal is based on High/Low levels which can be well defined and which will be much less affected by noise and interference than analog.
b) Many two-state electronic switches exist, making it easy for manufacturers to implement complex digital circuits. With VLSI (very-large-scale integration), more than 100 000 transistor switches can be arranged in one IC chip.
c) A digital signal, being a group of 1s and Øs, is easy to store, process and manipulate.

This ability to store, process and manipulate data gives the **digital system** great flexibility over its actions. The system can therefore be very versatile, limited only by the ingenuity of the program. An example comparing an analog type with a digital motor speed control system will illustrate this point.

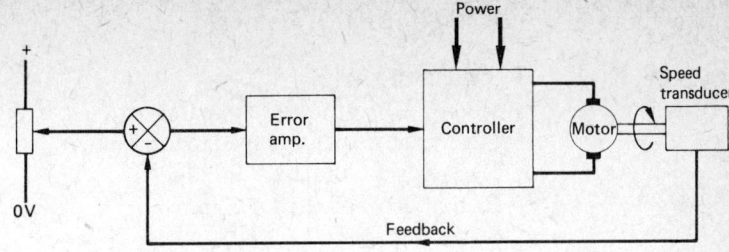

Fig. 1.13a Motor speed control: analog system

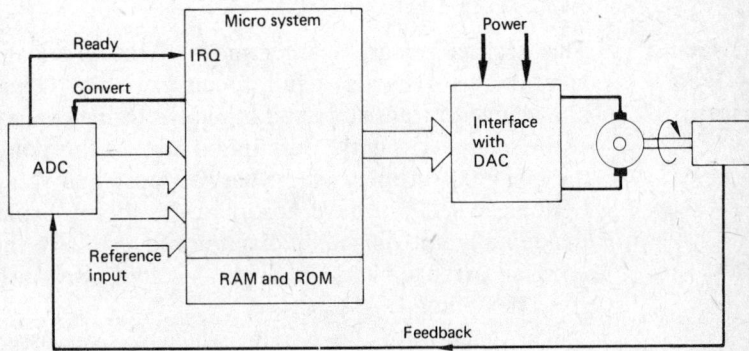

Fig. 1.13b Motor speed control: digital system

Shown in *fig. 1.13a* is the block diagram of the analog version. The required speed is set by the voltage from the wiper of the input potentiometer and this is compared with the output signal from the speed transducer. The comparator output, which will be the error or difference between required speed and actual speed, is amplified and fed to the controller. Consequently, power is applied to the motor to adjust its speed until the error is reduced to a low value. The system is relatively simple, will give accurate control of speed, and is able to respond to changing load conditions.

When constructed in digital form (*fig. 1.13b*), the increased complexity of the system is apparent. The required speed can be read from an input device such as a keyboard or it could be stored within the program. The output from the speed transducer (assumed analog) must be converted into a suitable digital word via the ADC. This digital word, proportional to the actual speed, can be compared with the reference word by the program, and a digital output dependent on the result of this comparison will be used via the DAC and interface to adjust the motor speed. If the digital system was required to do only this, then the additional complexity compared with the analog form might be considered unjustified. But the program gives a lot of flexibility and provides the means of doing many things with the speed of the motor. In addition, a table of values proportional to motor speed can be stored somewhere in RAM or ROM and these can be called up by the program as desired.

A program example might be:

a) Switch on motor
b) Run motor up to full speed at a steady rate with a ramp of 20 sec
c) Hold motor speed constant for 5 min
d) Reduce speed to 10% in 20 sec
e) Run at 10% speed for 2 min
f) Switch off.

Note that with the digital system there is an endless variety of speed control possibilities that can be pre-programmed.

1.4 An Interfacing Example—Oven Temperature Control

This section brings together most of the ideas and techniques previously mentioned for use in a full circuit example. Obviously, if you have little knowledge of transistors and other electronic components, an understanding of the finer points in the description may, at this point, be over ambitious; but the general principles will be easy to grasp and you can always return to this example when you have absorbed the theory explained later in this book. Incidentally, the full circuit diagram will also show the apparent complexity of a digital arrangement, although as stated previously the digital system has great flexibility.

The function of the system, which was discussed in section 1.1, can be stated with a brief specification:

Temperature range: +35°C to 100°C
Transducer (temperature sensor): thermistor type GL23 (RS 151–029)
Accuracy required: at least ±2.5°C
Sampling rate: every 15 sec
Heater: 500 watts resistive
Heater supply: 240 V 50 Hz single-phase a.c.
Control device: thyristor via bridge rectifier.

The system block diagram is shown in *fig. 1.14*. The input sensor is the thermistor, a device that exhibits a marked change of resistance with temperature. The change in resistance (the resistance of this thermistor *falls* as the temperature rises) can be converted into an analog voltage by supplying the thermistor from a constant-current source. Using Ohm's law:

$$V = IR$$

where V = output voltage
I = current
R = resistance.

When I is held constant

$$V \propto R$$

The output voltage, which will fall with increasing temperature, must then

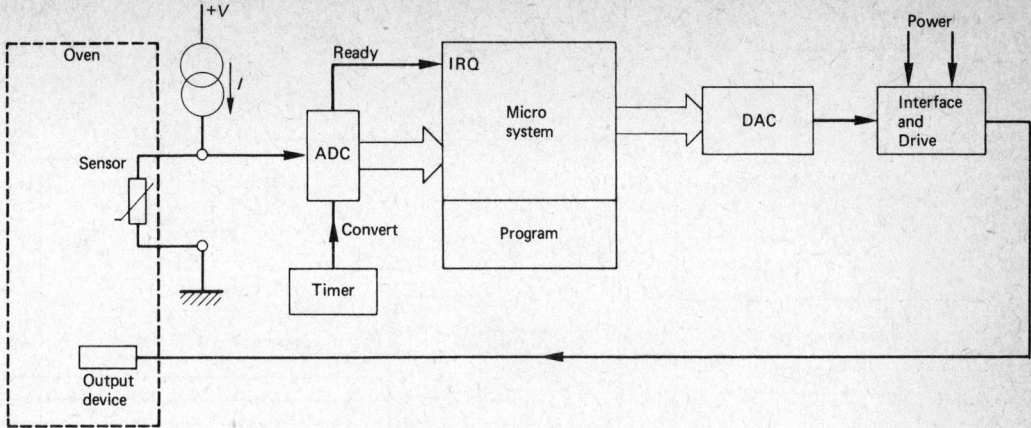

Fig. 1.14 Block diagram of proposed temperature-control system

be converted by an ADC into digital form before being applied to the micro system. Note that an external timer which pulses the ADC once every 15 sec has been shown. This hardware approach does not require the microprocessor to be idle while it waits to sample the ADC output. Instead, after a short interval following the CONVERT pulse from the timer, a READY signal is generated which interrupts the microprocessor and causes it to service the ADC.

On the output side of the micro system, the digital output must be converted into analog before being used to control the power delivered to the heater.

1   *Fig. 1.15* shows one way in which the **input** side of the system can be built up with actual components and we can look at the operation of this block by block.

The pnp transistor circuit formed around Tr_1 supplies the constant current to the thermistor. With the resistor values given, a fixed voltage of approximately 1 V is set up across R_3 and RV_1; and by varying RV_1 the value of this current flowing out of the transistor's collector can be set between 0.2 mA up to 1.5 mA. Suppose the current is set to 1.275 mA. At 20°C, the GL23 thermistor has a stated resistance of 2 kΩ and therefore the analog voltage will be 2.55 V. At 40°C, the resistance of the thermistor falls to about 1 kΩ and the voltage will drop to 1.27 V. At the maximum temperature of 100°C, the thermistor's resistance is just below 200 Ω giving an output voltage of 0.24 V. Note that the fall in voltage will not be linear and careful calibration of the system would have to be carried out to achieve the required accuracy.

The basic formula for finding the resistance of this type of thermistor is

$$R_2 = R_1 \, e^{B(1/T_2 \,-\, 1/T_1)}$$

where B = the characteristic temperature constant (K)
T = bead temperature (K)
R_1 = resistance of thermistor at temperature T_1
R_2 = resistance of thermistor at temperature T_2.

16 Practical Interface Circuits for Micros

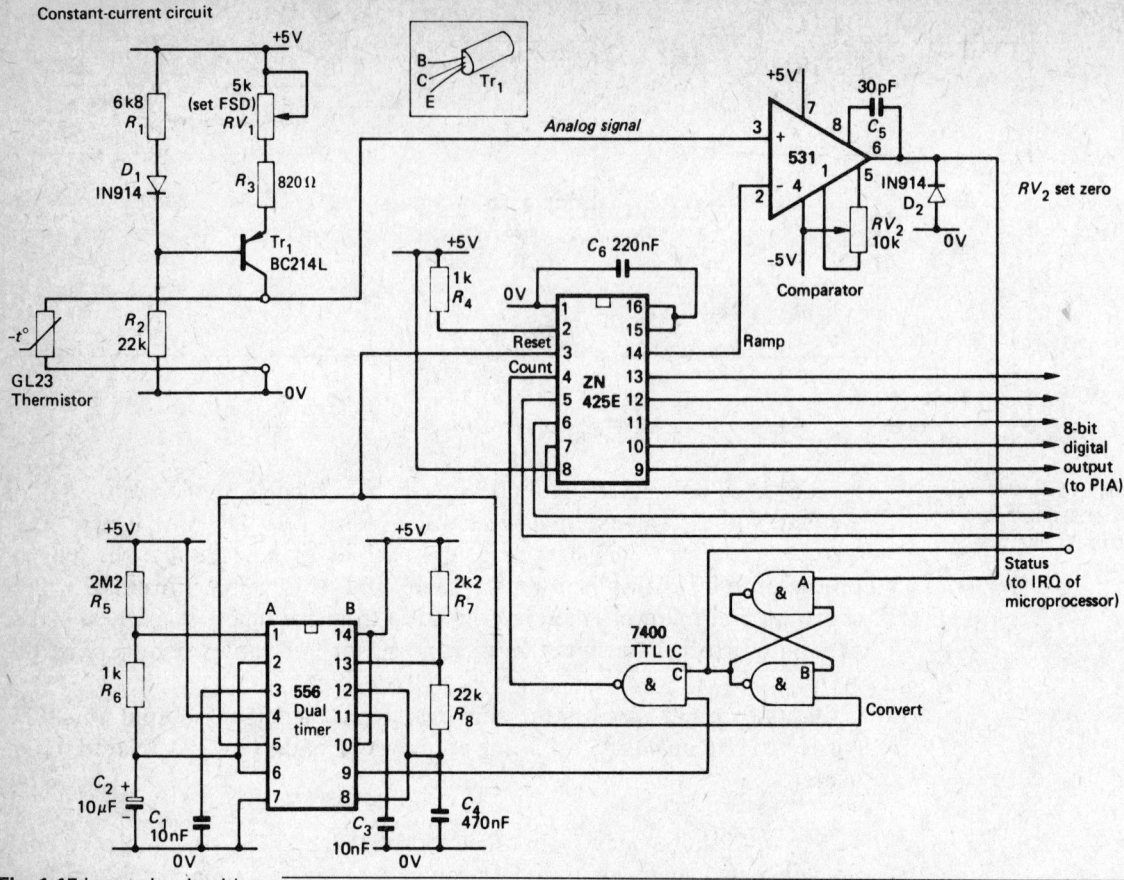

Fig. 1.15 Input circuit with ADC for oven temperature controller

Fig. 1.16 Thermistor resistance and output voltage against temperature

Temperature °C	Thermistor resistance Ω	Analog voltage V	Digital output	Hex. value
20	2 000	2.55	1111 1111	FF
25	1 665	2.123	1101 0100	D4
30	1 395	1.780	1011 0010	B2
35	1 175	1.50	1001 0110	96
40	995	1.268	0111 1111	7F
45	847	1.08	0110 1100	6C
50	725	0.925	0101 1101	5D
55	624	0.796	0101 0000	50
60	539	0.687	0100 0101	45
65	467	0.595	0011 1100	3C
70	407	0.519	0011 0100	34
75	356	0.454	0010 1110	2E
80	312	0.398	0010 1000	28
85	275	0.350	0010 0011	23
90	243	0.310	0001 1111	1F
95	216	0.280	0001 1100	1C
100	192	0.245	0001 1000	18

Both T_1 and T_2 are in Kelvin (°C + 273 = K).

Using this formula and given that $B = 3200$ for the thermistor specified, a table of resistance values against temperature can be drawn up as shown in *fig. 1.16*. Also shown in this table are the analog voltage levels for $I = 1.275$ mA and the rounded-up respective 8-bit digital outputs required from the ADC. Because the thermistor is a highly nonlinear device, the conversion values become more cramped at the top end of the temperature range, but it is possible for the system to distinguish between temperature settings of 95°C, 97.5°C and 100°C to give the required accuracy.

The analog-to-digital convertor is created using a ZN425E DAC, a comparator (531 op-amp), and some logic gates (7400). The action of this circuit is controlled by the timer (556 A side). The 15 second wait time between sampling is set by R_5 and C_1. The operation is more easily followed by studying the waveform diagram in *fig. 1.17*.

When the timer output goes low for the few milliseconds just after the main delay, the following actions take place:

a) The control bistable is set. Gate B output goes high.
b) Status rises high.
c) The internal counter in the ZN425E is reset and held.

At the end of the relatively short negative pulse from the timer, the reset is removed and clock pulses (556 B side) pass through gate C to the ZN425E. The 8-bit counter accumulates counts and inside the ZN425E the state of the counter is converted by a resistive ladder network (R-2R type) into a ramp

Fig. 1.17

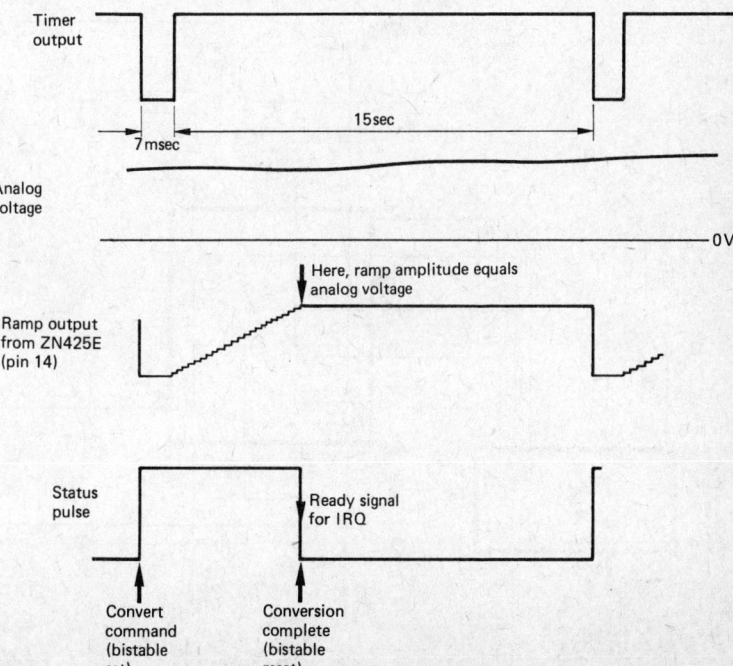

18 Practical Interface Circuits for Micros

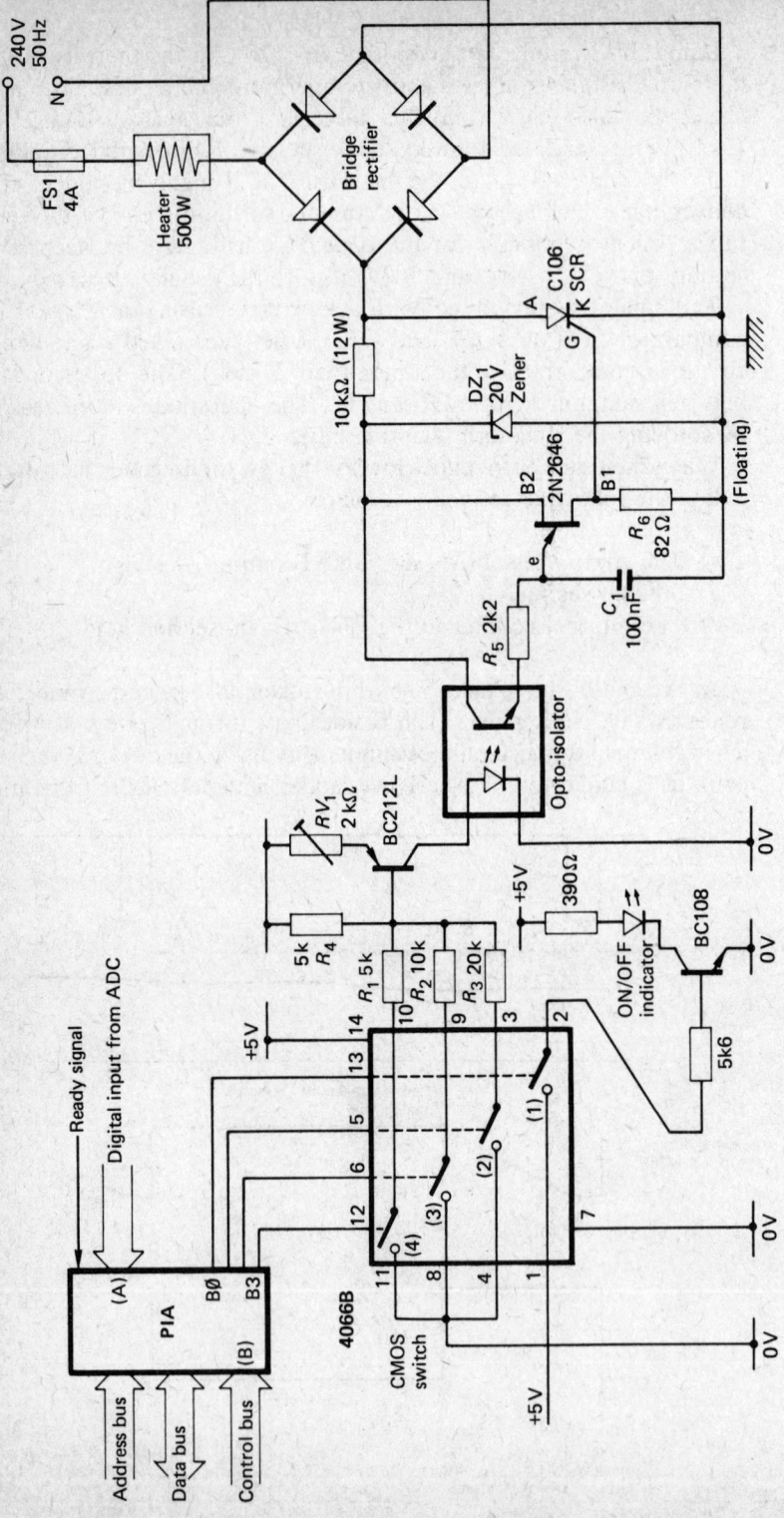

Fig. 1.18 Interface circuit to heater

output at pin 14. This ramp, which can have a maximum of 255 steps and a voltage level of 2.5 V, is then compared with the analog voltage from the sensor.

Suppose the temperature is about 40°C, giving an analog output from the thermistor circuit of 1.27 V. When the ramp from pin 14 of the ZN425E just exceeds 1.27 V, the output of the 531 comparator switches low and resets the bistable. This in turn prevents any further clock pulses being sent to the counter, and the state of the counter will be a digital word proportional to the analog voltage, in this case of 01111111 or $7F_{hex}$. At the same time that the bistable is reset, the STATUS level goes low, giving a READY signal to interrupt the microprocessor. When this occurs, the microprocessor completes its current instruction (assuming it is being used for another controlling job in addition to the oven heating) and then fetches the address of the routine for temperature control from a vectored address in memory. Having loaded the start address of the temperature control program into its program counter, it fetches and executes the first instruction. This could be, for example (using an M6800 processor),

    LOAD Accumulator A from PIA

where PIA is the **Parallel Interface Adaptor**. In other words, Load Accumulator A (a working 8-bit register in the 6800) with the digital word from the ADC. The program then causes the microprocessor to compare this digital input with a reference word held in memory. Following the result of this comparison, the microprocessor outputs a control word via the PIA to the heater interface. The PIA has an internal register which will store the control word during the time delay between ADC interrupts. This allows the microprocessor to return from the heater control subroutine and to carry on with any of its other tasks.

2 The interface circuit used for the **output** to the heater is shown in *fig. 1.18*. This again is just one example out of several that could be chosen. Only 4 bits are used for the interface control:

    Bit 0   ON/OFF indicator (bit 0 = 1 for ON)
    Bit 1⎫
    Bit 2⎬ Heater power level control
    Bit 3⎭

The 3 bits used for heater power control will give 8 possible levels, from 000 (heater off) up to 111 (full power), each of these steps being approximately 70 W. This is illustrated in the table in *fig. 1.19*.

The 4-bit output from the microprocessor via the PIA is used to operate the four electronic switches contained in a 4066B quad analog switch IC. These switches then connect a variable current via selected weighted resistors to the light-emitting diode in an opto-isolator. (Since the heater is connected to the 240 V a.c. mains supply, isolation is an essential feature.) In this way the current that flows through the LED is varied by the digital output from the microprocessor.

20 Practical Interface Circuits for Micros

Output from micro (hex.)	4-bit word binary (lower 4 bits of output)	Resulting power level (approx.)(W)
ØØ	ØØØØ	OFF
Ø3	ØØ11	70
Ø5	Ø1Ø1	140
Ø7	Ø111	210
Ø9	1ØØ1	280
ØB	1Ø11	350
ØD	11Ø1	420
ØF	1111	490

Fig. 1.19 Output word to heater interface

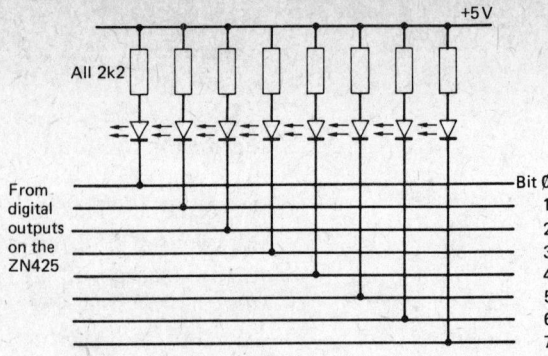

Fig. 1.20 Test circuit for digital output; LED lights when Ø present on the appropriate bit

For example, if the digital output is Ø3 (ØØ11), switches 1 and 2 will be on but switches 3 and 4 off. The current through the LED will be limited to its lowest value. At low levels of current the light intensity from the LED will also be low and only a small current will flow in the output transistor inside the opto-isolator. The timing capacitor C_1 is charged relatively slowly compared with the time period of one half-cycle of the mains waveform so that the unijunction gives an output pulse late in each half-cycle. The thyristor therefore conducts late in each half-cycle and only a small amount of power will be dissipated in the load. Increasing the LED intensity by outputting a larger word from the microprocessor, say ØB (1Ø11), will cause C_1 to be charged much more rapidly and the thyristor will be triggered on early in each half-cycle to give relatively high power dissipation in the load.

It is important to realise that the controlling action would probably be nonlinear; some adjustment to the values of R_1, R_2 and R_3 might be necessary to achieve more linear control. The circuit is an example of the phase control technique and as such can be used without much modification to control the speed of an a.c. motor or the brilliance of some lamps. In such cases it would then be necessary to change the input transducer and the timer sampling rate. For lamp brilliance control, a light-dependent resistor (photo-conductive cell type ORP12) would be most suitable. This could be fitted as a direct replacement for the thermistor and the value of the constant current would be adjusted to give maximum output voltage (2.5 V) at minimum brilliance level.

3   Before we move on to discuss the possible programming of the system it is worth noting that both the input interface and the output interface can be tested and set up independent of the micro system. In this way the various portions of the whole system can be proved before the software is written.

The degree of **testing** will depend upon the availability of suitable test equipment. Here we will assume the minimum of just one analog multimeter. This should have a resistance of at least 20 kΩ/V on d.c. ranges.

To *test for correct input interface and ADC operation* use the following type of procedure:

a) Measure the d.c. voltage across the thermistor. This should be approximately +2.5 V (assuming room temperature). Adjust RV_1 to set the voltage just below +2.5 V.
b) Heat up the thermistor slightly and note the change in voltage. The voltage should fall.
c) Check the timer output. Connect the meter on 10 V d.c. range between pin 5 of the 556 and ground. The meter should indicate about 4.2 V and every 15 sec or so there should be a short downward meter deflection.
d) Check status output. This should be a short-duration postive-going pulse occurring approximately every 15 seconds. Connect the meter on 10 V d.c. range between the status output and ground. The presence of this pulse indicates that the bistable is being set and reset correctly.
e) Check digital outputs from the ZN425. Each output will give a current of only 49 µA when in the logic 1 state, which is insufficient to drive an LED indicator without a buffer. However, the outputs will each "sink" 1.6 mA when in the logic Ø state. A small indicator unit can be made consisting of eight 2k2 resistors and LEDs for checking the complement of the digital output (see *fig. 1.20*). Test that the display changes as the thermistor is heated.

For *testing the output circuit operation*, the heater can be replaced temporarily by a 100 W lamp.
IT IS MOST IMPORTANT THAT SAFE PRACTICES ARE FOLLOWED IN WIRING AND TESTING THE OUTPUT CIRCUIT. THE MAINS VOLTAGE IS LETHAL.

a) Set all four inputs of the quad switch to Ø and check that both the ON/OFF indicator *and* the lamp load are off.
b) Set the digital input to ØØØ1. The ON/OFF indicator should be on but the lamp load off.
c) Set the digital input to 1111 and adjust RV_1 so that lamp brilliance is just at maximum.
d) Vary the digital input as follows and check output:

    ØØ11, Ø1Ø1, Ø111, 1ØØ1, 1Ø11, 11Ø1 and 1111

There should be a reasonably linear change in lamp brilliance from minimum to maximum. If necessary make further adjustment to RV_1 and/or the values of R_1, R_2 and R_3 to achieve the most satisfactory control.

4 Once satisfied that the input and output circuits operate correctly, the task of writing the software can begin. In this example, the **program** will be assumed to be ultimately written for a Motorola 6800 based system with a user PIA at address $8ØØ4. At this stage it is not intended to produce a full assembly language program but rather to develop the flowchart and to discuss the programming problems set up by the oven temperature control system. It is also worth noting that, once the main task of creating a suitable flowchart has been successfully completed, the actual software can be easily produced for any other 8-bit microprocessor systems.

To get a system fully working, the tasks are as follows:

a) Initialise the PIA.
b) Initialise the microprocessor by clearing the interrupt mask bit (this allows the micro to be interrupted); setting the stack pointer; and loading the vectored address.
c) Construct a suitable flowchart.
d) Write an assembly language and machine code program.
e) Run and test the program.

Without going into too much detail of the PIA at this stage, the process of **initialisation** (referred to as "initialising" or "configuring") is the method using software by which the PIA is set up to suit a particular requirement. For our system the PIA must be set up or configured to give:

1 All 8 bits on the A side as inputs.
2 The lower 4 bits on the B side as outputs.
3 The CA1 interrupt input to be triggered from the negative edge of the status signal.
4 The interrupt mask bit cleared.

This is shown in diagram form in *fig. 1.21* and the actual program construction for the Motorola PIA initialisation is fully discussed in Chapter 6.

During operation of the system, the following sequence of events takes place:

a) The external timer gives a pulse that initiates the conversion and, when the ADC has completed the conversion, the STATUS signal goes low to give a READY signal to interrupt the microprocessor.
b) The microprocessor will execute the current instruction (i.e. any other program instruction it is working on at that time); it will then save the contents of the internal registers on the Stack so that at the end of the temperature control routine it can continue with its other tasks, and then load the address of the "Temperature control" routine from the interrupt vectored addresses FFF8 and FFF9. The microprocessor will do this as long as the interrupt mask bit has been previously cleared (instruction CLI for the 6800). Note also that the stack pointer register would have been loaded with an address to define the "stack" area in RAM and that the "Temperature control" routine start address must be held in locations FFF8 (high byte) and FFF9 (low byte).
c) The microprocessor will then begin the temperature control routine. The flowchart for this control action is shown in *fig. 1.22*. For this it is assumed that a series of hex. numbers proportional to oven temperatures in the range +20°C to +100°C in 2.5°C steps are held in a portion of RAM or ROM. This is referred to as a look-up table, a section of which is shown in *fig. 1.23*. A fairly complete set of values must be held because of the non-linearity of the thermistor.

Systems 23

Fig. 1.21 Flowchart for initialisation of PIA, interrupt mask bit, and stack pointer

Fig. 1.22 Flow chart for heater control program

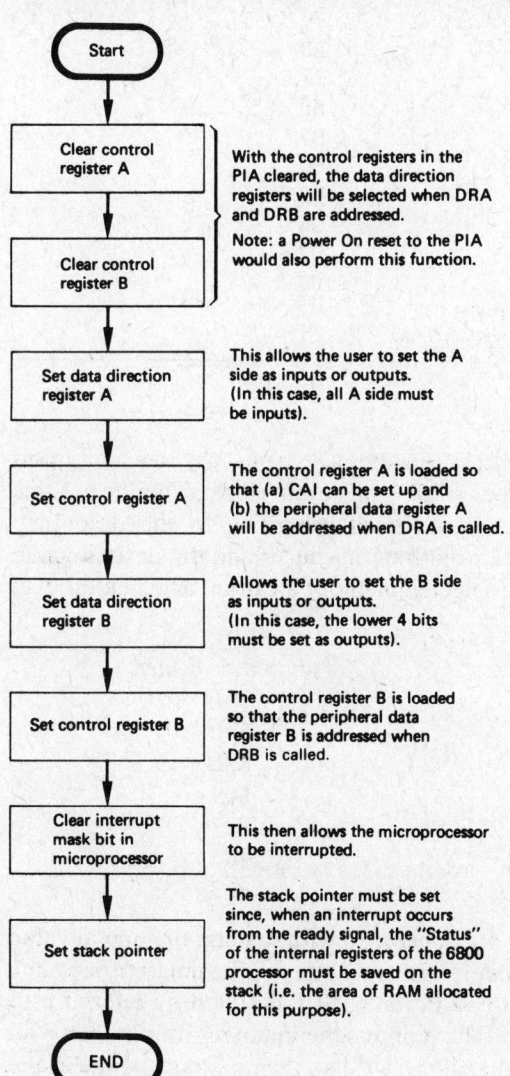

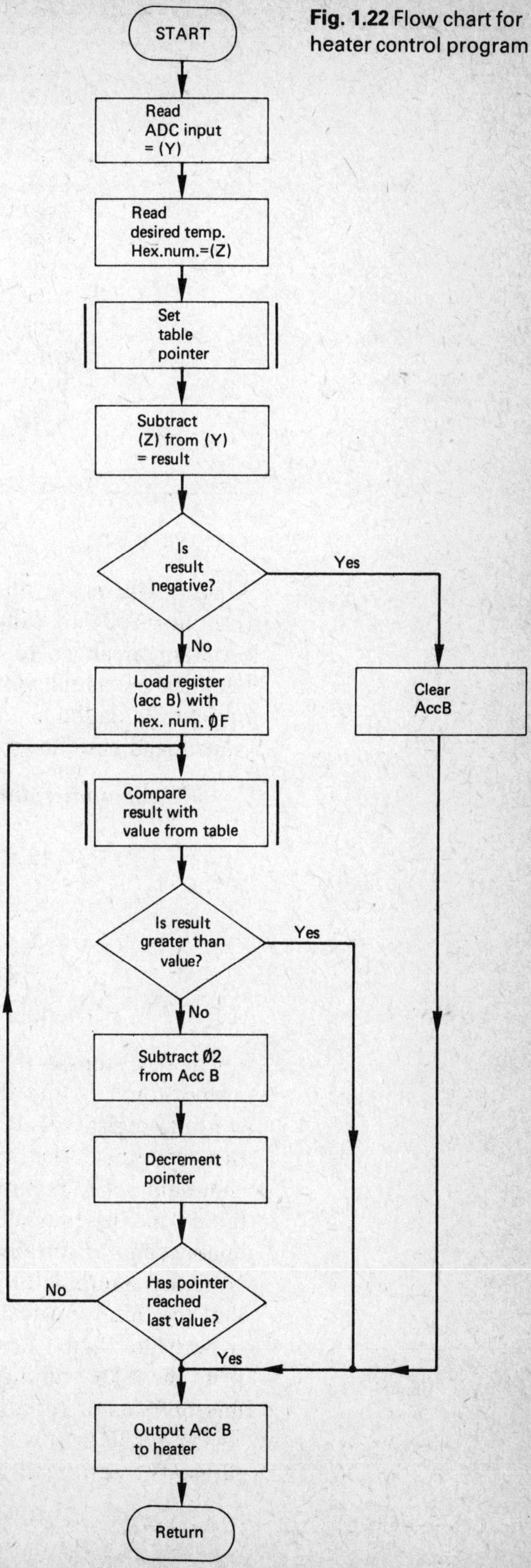

Fig. 1.23 Section of look-up table

	Address	Content	Equivalent temperature °C
	Ø1ØB	31	72.5
Pointer→	Ø1ØA	2E	75 ←
	Ø109	2B	77.5
	Ø108	28	80 ←
	Ø107	25	82.5
	Ø106	23	85
	Ø105	2Ø	87.5
	Ø104	1F	90 ←
	Ø103	1E	92.5
	Ø102	1C	95
	Ø101	1A	97.5
	Ø1ØØ	18	100
		Part of RAM or ROM	

You will recall that the heater can be set from an OFF condition (output = $ØØ) to fully ON (output = $ØF) in seven steps. Therefore, this program arranges to adjust the heater level according to the difference between the actual value recorded from the transducer and the desired value previously loaded in and held in a selected memory location. The heater level is arranged as follows:

Temperature difference °C	Output to heater
>15	$ØF
>12.5	$ØD
>10	$ØB
>7.5	$Ø9
>5	$Ø7
>2.5	$Ø5
equal	$Ø3

The first steps in the flowchart are concerned with loading the digital value proportional to the desired temperature into one of the microprocessor's working registers (say accumulator B in the 6800), and then by reference to this value to set the table pointer. The pointer, the index register in the 6800, is initially set to point a memory location 15°C less than that hex. value in the table which is equivalent to the desired temperature. Setting the pointer will enable comparisons to be made between the actual and desired temperatures. Note that a number of program steps would be required for this operation and that the difference values are not linear.

Next the digital word proportional to actual temperature must be loaded from the ADC via the PIA data register (A side) into another register in the microprocessor (accumulator A for example), and a subtraction of the two registers (SUBA) will give the difference in hex. between the two temperatures. The result will be held in accumulator A.

If the result is negative, indicating that the oven temperature is already higher than the desired value, the program will branch to output $ØØ via the PIA (B side) to the interface. The heater will be turned off.

Assuming, however, that the oven temperature is low, the program will continue by first loading a register (accumulator B) with the hex. number used for maximum heater power, i.e. $ØF, and then successively compare the result (difference between temperatures held in accumulator A) with the hex. difference values for 5°C, then 12.5°C, and so on, generated from the table. If, each time a comparison is made, the result is found greater than the value generated from the table, then $Ø2 is subtracted from accumulator B and the table pointer is decremented.

Take an example where the desired temperature is +90°C, which is equivalent to the hex. number $1F, and the actual temperature is 80°C equivalent to the hex. number $28. The pointer will initially be set to memory location $Ø1ØA (see *fig. 1.23*) where the hex. number held ($2E) is equivalent to 15°C less than the desired temperature. On the first pass, the "result" will be less than the difference between 75°C ($2E) and 90°C ($1F) indicating that the temperature difference is closer than 15°C. Accumulator B will have $Ø2 taken from it (it will then hold $ØD), and the index register will be decremented causing the pointer to point to memory location $Ø1Ø9. On the second pass, the result will still not be greater than the difference value generated from the table, and therefore the content of accumulator B will be reduced to $ØB and the index register decremented so that it points to memory location $Ø1Ø8. However, on the third pass, the result will not be greater than the new difference value generated from the table ($28 − $1F is greater than $25 − $1F); then a branch occurs so that the contents of accumulator B are outputted to the heater interface via the PIA. The heater will receive about 5/7 power.

The heater control routine finishes and the return from interrupt instruction (RTI) causes the microprocessor to restore the internal registers, namely the condition code register, accumulators B and A, the index register, and the program counter, to their previous state prior to the interrupt by pulling the values from the stack. The microprocessor can continue its other tasks during the 15 seconds before the temperature sensor generates the next interrupt.

More detail of the 6800 micro will be given later.

2 Useful Electronic Devices and Circuits

2.1 Electronic Component Ratings

There are several electronic components and circuits which can be put to good use in interfacing a microcomputer for control purposes, and the following sections contain details and explanatory notes on a reasonable selection of analog types. The treatment is naturally not in depth, but more a review of the main features and of how to make practical use of each device.

As well as understanding how devices operate, it is very important to appreciate how they are rated. Any component is usually well described in the manufacturer's data sheet which will give a mechanical outline, connection details, plus important electrical characteristics and parameters. These can be a very valuable guide and should be consulted if possible when a device new to you is to be used. Before connecting in any component, always check that in your application the absolute maximum values of voltage, current and power are not going to be exceeded. This is particularly true for sensitive and expensive ICs but applies also to humble components like resistors and diodes. Operating a component well within its rated maximum values will prolong its life.

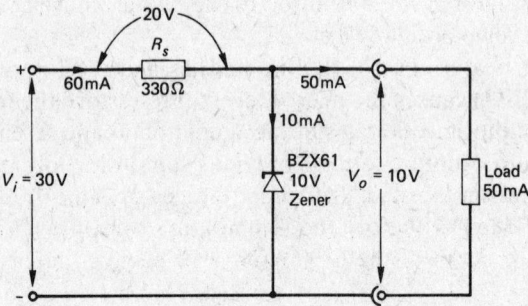

Fig. 2.1 Circuit of simple voltage regulator to illustrate calculations of power dissipation

Take the simple example of a shunt voltage regulator as shown in *fig. 2.1*. The zener or voltage regulating diode holds the output voltage reasonably constant irrespective of changes in the input voltage or the load current. To calculate the maximum power loss of the zener diode, imagine that the load is removed; all the load current (50 mA) will then flow through the diode causing the zener current to rise to 60 mA. That is

$$I_{Z\max} = 60 \text{ mA}$$

Since $V_Z = 10$ V,

$$P_{Z\max} = V_Z I_Z \simeq 600 \text{ mW}$$

The BZX61 zener diode has a power rating of 1.3 W at an ambient temperature of 25°C and is then a suitable choice for this circuit. The power handling capability of the diode decreases by 8.7 mW/°C and therefore even at a high temperature of, say, 45°C its power rating at 1.126 W [1300 − 8.7 × 20 = 1126] is well in excess of the maximum dissipation in the circuit.

The series resistor will dissipate the following total power:

$$P_{R(\text{tot})} = V_R I_R$$

where $I_R = I_Z + I_L = 60\,\text{mA}$
and $V_R = V_i - V_o = 20\,\text{V}$

$\therefore\ P_{R(\text{tot})} = 2 \times 60 \times 10^{-3} = 1.2\,\text{W}$

Without first having checked the actual power dissipated by the resistor it might be tempting to fit a 1 W type or indeed any 330 Ω resistor to hand. This would then run very hot, or worse still, burn out. The type to put into this circuit is a general-purpose wire-wound resistor with a power rating of, say, 3 W; this would then easily dissipate the normal power and be able to withstand the overload if either the zener or the load became short circuit. Under these worst-case conditions the current through the series resistor would be

$$I_{\text{sc}} = V_i / R_S = 30/330 = 91\,\text{mA}$$

$\therefore\ $ Worst-case power dissipation of R_S is

$$V_i \times I_{\text{sc}} = 30 \times 91 \times 10^{-3} = 2.73\,\text{W}$$

In practice, simple calculations like these can be made very quickly by estimating the maximum value of voltage to be expected across a device and multiplying this by the maximum current that will possibly flow through it.

2.2 Semiconductors—Diodes, Transistors and FETs

Practically the whole of modern electronics depends upon the special properties of the semiconductors such as germanium and silicon. A substance like silicon has, in the pure state, the sort of atomic structure which makes it neither a good conductor nor a good insulator, but by introducing small quantities of impurities into the pure silicon it is possible to modify the conducting properties and then to create devices such as diodes, transistors, and integrated circuits. The impurities are forced into selected regions of a silicon slice during the manufacturing process; this process is called *doping*. Basically, two types of impurity are used which convert the silicon into either n-type or p-type (*fig. 2.2*). The **n-type** has an excess of almost free electrons which are not bound tightly to atoms; these electrons can be used to carry electric current. The opposite type of material, **p-type**, has "holes" (a deficiency of electrons) which will also carry current but in the opposite direction to electrons.

One of the important building blocks of semiconductor components is the **pn junction**. One pn junction is used for a diode or rectifier; two junctions

28 Practical Interface Circuits for Micros

Fig. 2.2a n-type

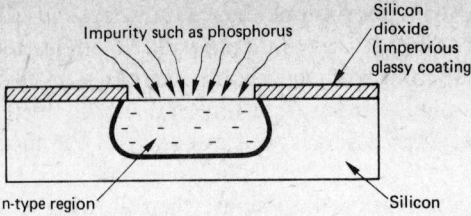

Fig. 2.2b p-type

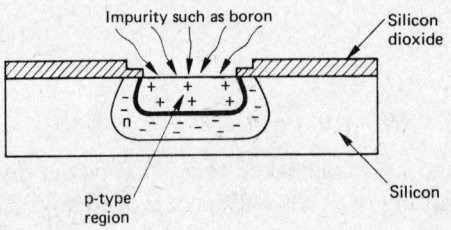

Fig. 2.3 Formation of pn junction

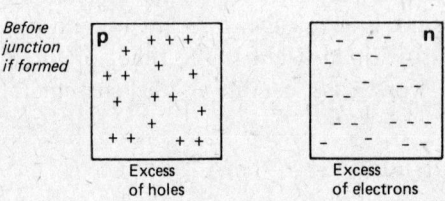

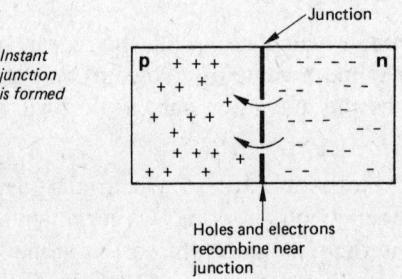

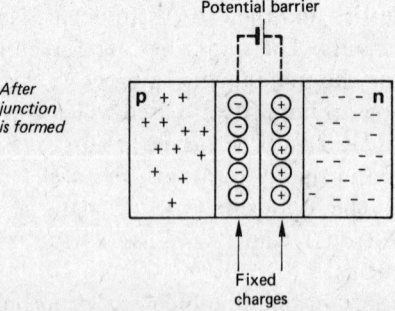

make a bipolar transistor; and three are used for a device such as a thyristor. A pn junction is shown in *fig. 2.3*. At the instant a junction is formed (during manufacture), the electrons and holes very near the junction recombine. But when an electron leaves the n material, the atom to which it was weakly bound becomes a fixed positive charge, and similarly when a hole in the p side is filled by an electron the atom where this occurs becomes a fixed negative charge. Thus, very rapidly a potential barrier is set up at the junction which prevents any further passage of charge carriers across the junction. The region just near the junction is then empty of charge carriers and is called the *depletion region*.

The pn junction, which then makes a diode, has the special property of conducting in one direction only; this is shown in *fig. 2.4*. If a *forward bias* of about 0.6 V is applied, making the p-type positive with respect to the n-type, the potential barrier at the junction is overcome and charge carriers can cross the junction to make up a current flow. If the battery is reversed as in *fig. 2.4b*, the potential barrier at the junction is increased and the depletion region gets wider. Hardly any current flow can then take place.

Diodes come in all shapes and sizes, depending mostly on how much current is going to be passed through them. A good general-purpose signal diode is the IN4148 or IN914, which can pass a forward current of 75 mA (average) and withstand a reverse voltage of 75 V. Rectifier diodes, on the other hand, are usually designed to pass reasonably high currents and there are types that have a rating of 40 A or more. For medium power rectifier applications, the IN5400 range (3 A forward current rating) is a good choice.

Some basic applications of diodes are shown in *fig. 2.5*, all of them making use of the property of a diode to conduct one way only.

Fig. 2.4 Forward and reverse bias

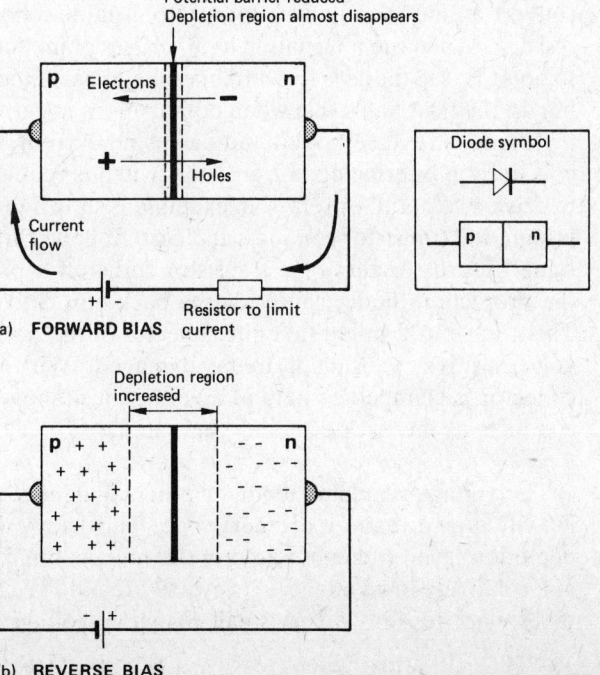

Fig. 2.5 Some applications of semiconductor diodes

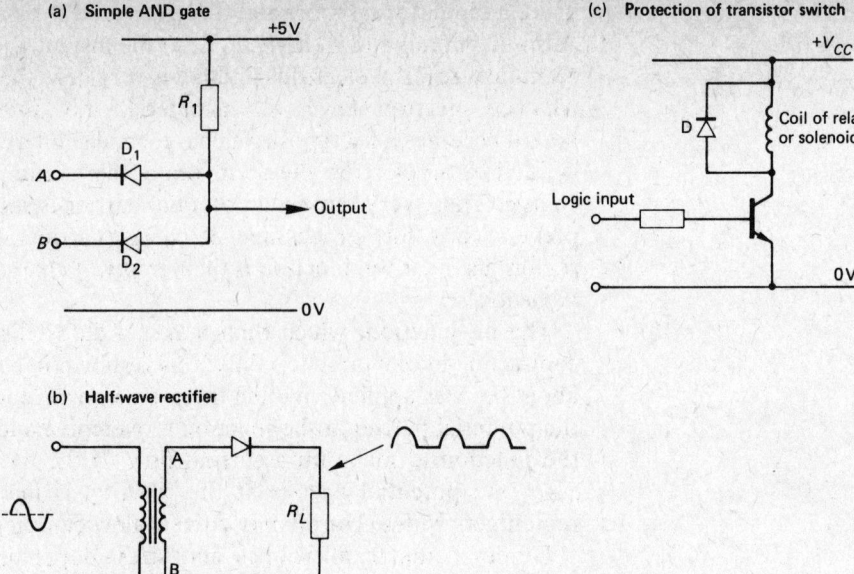

a In (a), two diodes and a resistor are used to make an AND gate. If both inputs A and B are high at, say, +5 V, then both diodes will have no forward bias and will not conduct. The output level will be high. However if either input is taken low to, say, 0 V, the diode at that input will conduct and the output will fall to approximately +0.6 V.

b Circuit (b) shows the basic half-wave rectifier, where the diode is used to convert an alternating waveform into a unidirectional signal across the load resistor. When the alternating input causes point A to be positive with respect to point B, the diode is forward biased and passes current through to the load; but on the next half-cycle when point A goes negative with respect to point B, the diode is reverse biased and passes no current.

c Often in interfacing, a transistor switch is required to buffer a logic signal to drive a solenoid or relay; an example is shown in (c). When the logic signal is high, the transistor conducts and current flows through the coil. As the logic signal returns to zero, the transistor turns off rapidly which would, without the protection diode, cause a large back e.m.f. to be generated by the coil. This back e.m.f. might take the collector of the transistor several hundreds of volts positive, causing it to be damaged. With the diode in circuit, the collector is clamped or held at a voltage just above $+V_{CC}$ because the diode conducts as the back e.m.f. is generated.

The **voltage regulator diode** or **zener diode** is a diode that has been more heavily doped than an ordinary type. This heavy doping results in a narrow depletion layer and consequently the reverse breakdown of the diode occurs at a relatively low voltage. At reverse breakdown, a rapid increase in current takes place for only a very small change in voltage across the diode, and it is

Fig. 2.6 Types of transistor

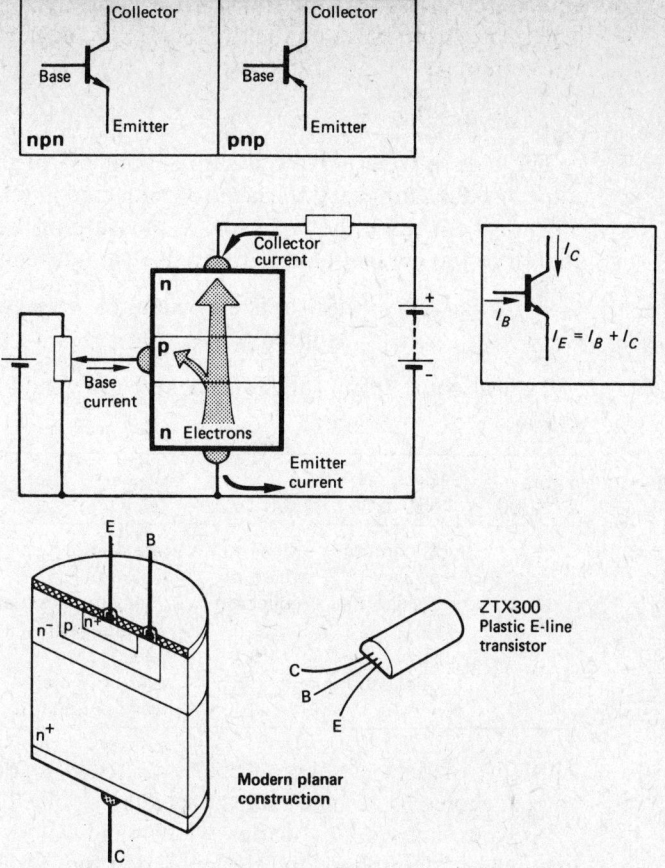

this characteristic which is used in circuits to provide a constant voltage. A circuit example has already been illustrated in *fig. 2.1*. Zener diodes are available in a wide voltage range, usually with a tolerance of ±5% of the value quoted, and with power ratings from 500 mW (BZY88 series) through a few watts (IN5333 with 5 W rating) up to 30 W or higher (BZY93 has 20 W rating).

The **bipolar junction transistor** has two pn junctions which give either an npn or pnp structure. The symbols and the npn structure (since this is the more common type) are shown in *fig. 2.6*. For correct operation, the silicon transistor requires a small forward bias voltage of about +0.6 V between base and emitter to overcome the potential barrier set up at the junction. Note that the collector/base junction is reverse biased. Electrons will then flow from the emitter into the base, but because of the doping levels there are few holes in the base available for recombination with these electrons. The majority of the electrons therefore spread out through the base region towards the collector until they reach the depletion region of the collector/base junction. Here they are swept up and "collected" by the positive field. Naturally this action is much more rapid than as described. The flow of currents, also given in *fig. 2.6*, shows that the transistor is basically a current-controlled device. The relationship between the three currents, neglecting any tiny leakage, can be written as

$$I_E = I_C + I_B$$

The base current is typically only about 1% of the emitter current, which indicates that, for every 100 electrons injected into the base from the emitter, 99% of them reach the collector. A measure of the quality of a transistor is therefore the **current gain** (written as h_{FE}) between base and collector.

$$h_{FE} = I_C/I_B \quad \text{for a fixed value of voltage between collector and emitter.}$$

Typical short form data for a transistor would be (at 25°C ambient) as follows:

Type	P_T	I_C	V_{CEO}	h_{FE}	f_T
ZTX300	300 mW	500 mA	25 V	50–300	150 MHz
	Maximum power dissipation ($V_{CE} \times I_C$ for your application).	Maximum value of collector current.	Maximum value of collector/emitter voltage with base open circuit (worst-case condition).	Current gain.	Frequency at which current gain falls to unity.

This sort of data enables comparisons to be made between types and the correct choice to be made for a particular application.

As you can see, h_{FE} varies widely and values as high as 900 are not uncommon. To understand the operation more fully, take the example of a transistor used as a switch to drive a relay as in *fig. 2.7*. When the logic signal is ∅, there is no forward bias to the transistor's base/emitter junction and no current flows in the collector. Both the transistor and the relay will be off. When the logic signal goes to 1 (+3 V), a base current of about 1.5 mA flows which turns the transistor fully on and the relay is energised. To ensure that the transistor is fully on, a ratio of only 10:1 is assumed between the collector and base currents. This is standard practice in switching circuits, but it is worth pointing out that the circuit would probably operate satisfactorily with a 20:1 ratio, that is with the base resistor increased in value to 3k3 ohms. Even so, if the logic unit supplying the base current, i.e. acting as a source to the base, can only be lightly loaded (some logic ICs can only source 40 μA while in the logic 1 state), a Darlington transistor or a power FET would have to be used in place of the transistor. Both these are discussed later.

The first example showed a transistor being used in a nonlinear manner as a switch. This is a vital action in logic integrated circuits where hundreds or thousands of transistor switches are used to make up a complete logic function. On the other hand, the bipolar transistor can be used in a wide variety of linear-type applications where its excellent amplifying properties, that is the current gain between its collector and base, are used. An easily

Fig. 2.7 Applications of a transistor as a switch

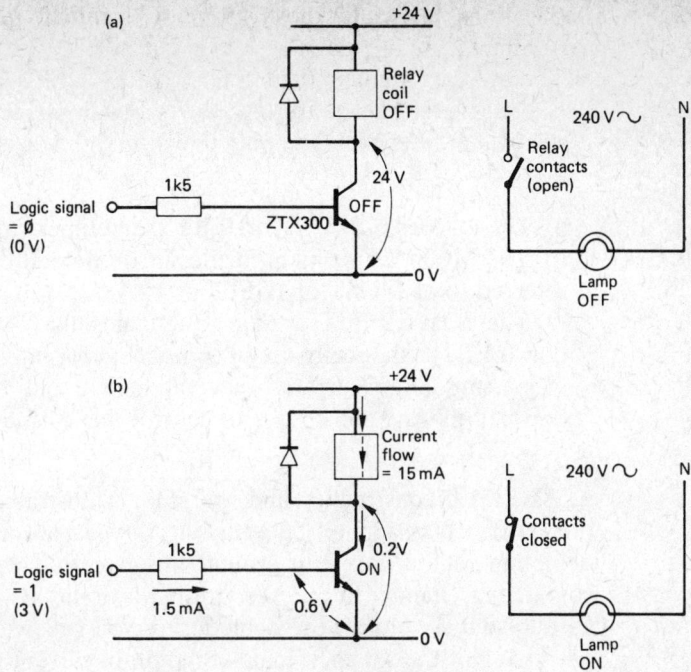

Fig. 2.8 A linear application of a transistor: simple series regulator

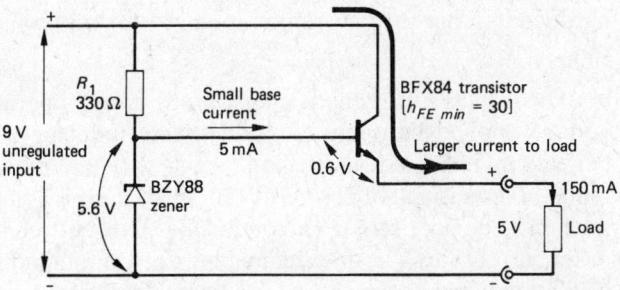

followed example is shown in *fig. 2.8* of a simple power supply regulator, where R_1 the 330 Ω resistor supplies current to both the zener diode and the base of the transistor. The base voltage is held constant by the zener at 5.6 V and the small base current of about 5 mA is amplified by the transistor to supply the much larger load current. Note that there is a voltage of about 4 V between the collector and emitter of the transistor which means that it dissipates about 600 mW. For this reason the BFX84 should be mounted on a small heat sink; this will prevent too high a rise of temperature at the collector junction.

The other type of amplifier, the **field effect transistor** or FET, does not require an input current for operation. With a FET, the current flowing through it is controlled by its input voltage. This type of action, which differs from the current-controlled bipolar transistor, is a distinct advantage in some applications, especially those where the available power to drive a device is very limited.

There are a number of different types of FET and confusion can arise because several names are used for the same device. The commonly used types, which have their symbols shown in *fig. 2.9*, are

a) The JUNCTION FET or JFET (sometimes called a JUGFET).
b) The MOSFET or metal oxide silicon field effect transistor (sometimes referred to as MOST or IGFET).
c) The POWER FET, with special names like VMOS and HEXFET. The power FET is the really useful component for interfacing since it is capable of passing large currents (several amps) and requires only a voltage between its gate and source to control this current.

a The JFET construction and operation is illustrated in *fig. 2.10*. Because of the reverse bias applied between the p-type gate and the n-type source, a depletion region is set up around the gate that restricts the width of the conducting channel between source and drain. By increasing the negative gate-to-source voltage, the depletion region gets wider and causes a further narrowing of the channel, making the drain current fall. If a sufficiently high negative voltage is applied (usually a few volts), the channel is cut off altogether and the drain current ceases. The important point to note is that hardly any input gate current is required for this controlling action.

b The MOSFET (enhancement mode type) operates by the controlling voltage applied between gate and source inducing a channel in the silicon beneath an insulated gate region (see *fig. 2.11*). Note that this is the opposite mode of operation to that of a JFET. In a MOSFET, a positive voltage (for an n-channel device) sets up a conducting path between source and drain and drain current flows in this channel between drain and source. Because of the insulation between the gate and the actual body of the device, a MOSFET has an even higher input resistance than a JFET.

c The power FETs such as VMOS and HEXFETs are enhancement mode MOSFETs which have the capability of passing very large currents. Suppose a 10 W rated solenoid (coil-operated lever) has to be controlled from a CMOS logic chip; the output of the CMOS logic, which must only be lightly loaded, can be connected directly to the gate of a VMOS power FET as shown in *fig. 2.12*. When the logic output is high at +9 V, the VN 46AF VFET is forced into conduction since the positive gate-to-source voltage creates the conducting channel between drain and source. A current of about 400 mA is passed through the solenoid coil, causing it to operate. When the logic signal returns to a low value, the VFET turns off and the solenoid current reduces to zero. Note the use of the protection diode wired across the solenoid coil to eliminate the turn-off "spike" which would damage the VFET.

Useful Electronic Devices and Circuits 35

(a) JFET

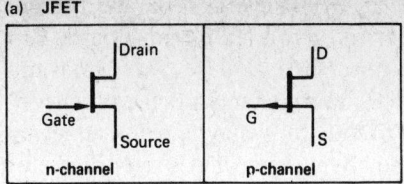

(b) MOSFET

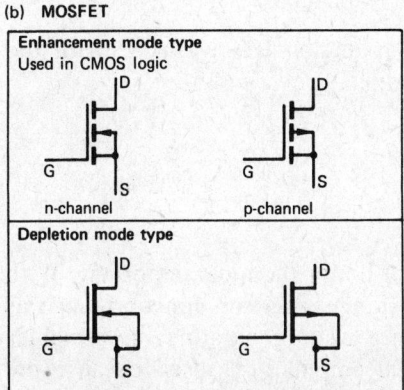

(c) Power FET — same symbol as the Enhancement mode MOSFET

Fig. 2.9 Field effect transistors

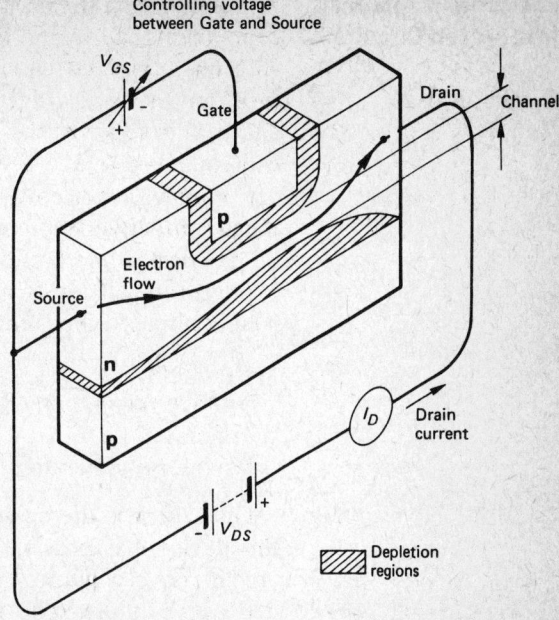

Fig. 2.10 Simplified view of construction and operation of an n-channel JFET

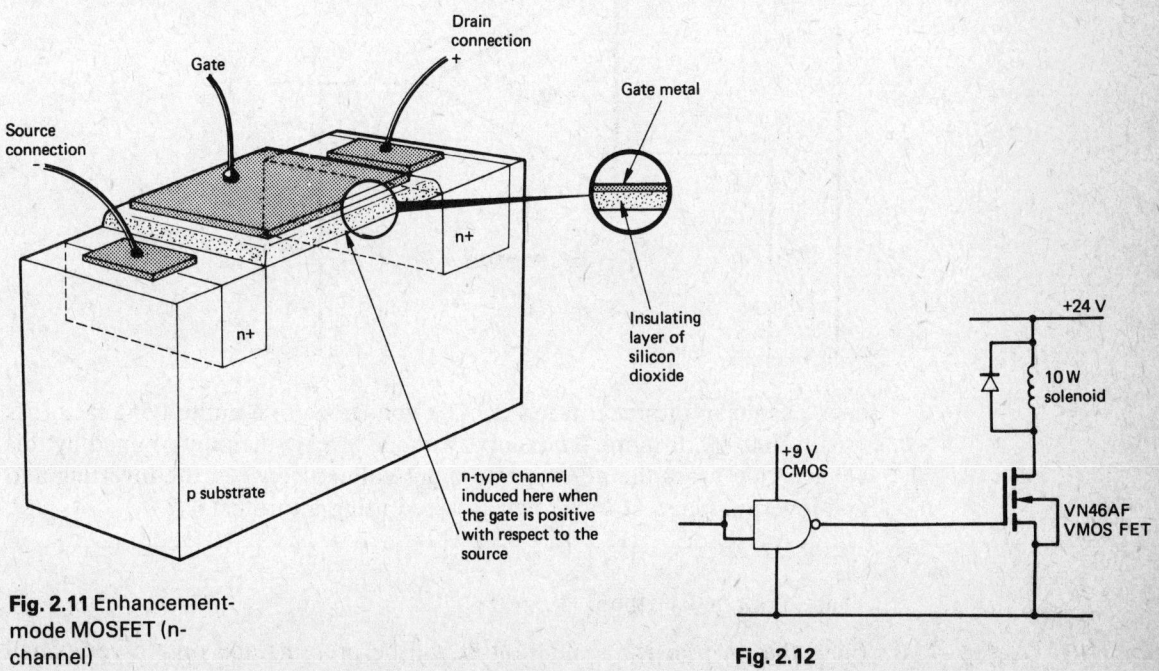

Fig. 2.11 Enhancement-mode MOSFET (n-channel)

Fig. 2.12

2.3 Linear/Analog Integrated Circuits

A working circuit can, of course, be built up from several separate discrete components, all linked together on a tag strip or printed circuit board. But with an integrated circuit or IC, all the necessary diodes, transistors and resistors for a particular function are "fixed" into one small piece of silicon, and it only requires the addition of power leads plus maybe a few external components to make a working circuit. The advantages of ICs over discretes are, mainly, lower cost, smaller size, and better reliability, but there is also the point that very complex circuits can be created in IC form that would not be practical using discretes.

There are all kinds of specialised linear/analog ICs available, with the commonly used types for our purposes being:

Op-amps
Voltage regulators
Timers
Waveform generators.

Of these it is the **op-amp** that is without doubt the most important. With this device it is possible to make a wide range of circuit functions such as amplifiers, oscillators, comparators, square wave generators, monostables and active filters. The name "operational amplifier" is used because the circuit was originally developed for analog computer work; it is basically a very-high-gain differential direct-coupled amplifier circuit (*fig. 2.13*). What

Fig. 2.13 Op-amp

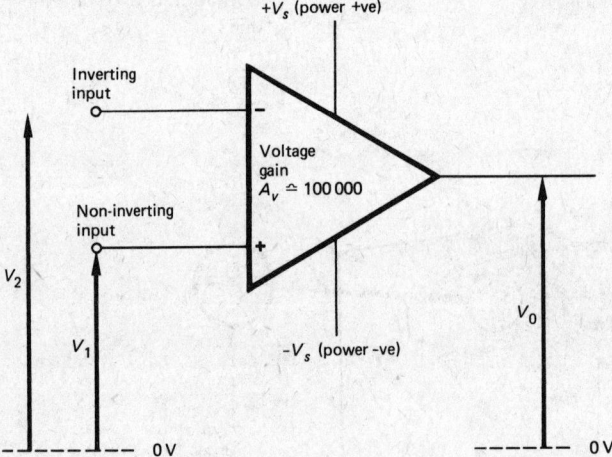

does this mean in practical terms? The voltage gain, a multiplying factor, is usually 100 000 or more. The output voltage level is then determined by this voltage gain times the *difference* in input voltage between the inverting and non-inverting pins. This can be expressed using a formula:

$$V_o = A_v (V_1 - V_2)$$

Thus if $V_1 = V_2$ then $V_o = 0$

(note that in practice a small offset will be present) and only a very small "differential" between the input voltages is required to give a large output voltage.

For example, if $V_1 = +0.1\,\text{mV}$ and $V_2 = 0$, then

$$V_o = 100\,000 \times (0.1\,\text{mV}) = +10\,\text{V}$$

Similarly if $V_1 = +0.1\,\text{mV}$ and $V_2 = 0.2\,\text{mV}$, then

$$V_o = 100\,000 \times (0.1\,\text{mV} - 0.2\,\text{mV}) = -10\,\text{V}$$

Since the amplifier circuit is directly coupled, in other words there are no coupling capacitors in the circuit, it will respond to d.c. levels at its inputs as well as varying signals.

However, an op-amp will have an upper frequency limit and the parameter which is concerned with the speed response of the device is called *slew-rate*. This is measured in volts per microsecond, and is the speed with which the output can change when a sudden step-like input is applied. The faster the speed, the better will the op-amp be able to follow high frequency input signals. Slew-rate can vary from as low as $0.4\,\text{V}/\mu\text{sec}$ (type 301A) to $35\,\text{V}/\mu\text{sec}$ (type 531) or higher. High speed is really essential in comparator applications.

Offset, which has been previously mentioned, is caused by small differences in the properties of the input components inside the IC. It is defined as the input voltage that would have to be applied between the two input pins to adjust the output to exactly zero volts. In many applications the tiny offset does not cause problems and where it does most op-amps are provided with pins to which a potentiometer can be connected to give what is called *offset null facility*.

We shall concentrate on applications of the op-amp which are particularly useful in interfacing and this should then show how the devices can be used, and also more about the way in which they operate.

Fig 2.14a Op-amp used as a buffer (unity gain voltage follower)

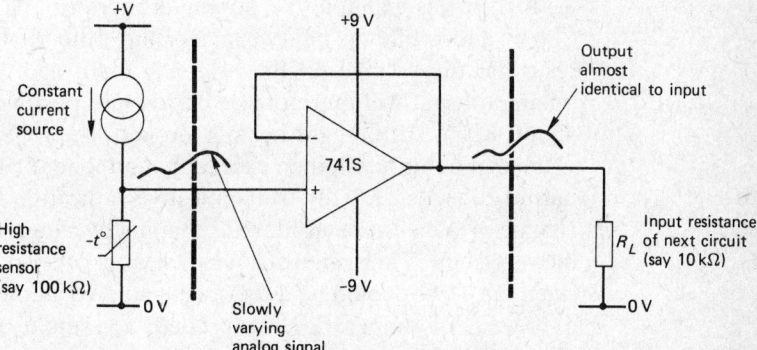

In *fig. 2.14a*, an op-amp is wired up as a **unity gain voltage follower** to act as a buffer between a relatively high resistance sensor and the load presented by some circuit. The output of the op-amp is connected directly back to its inverting input, and the signal from the sensor is connected to the non-inverting input. This connection method gives not only near unity voltage gain but also very high input resistance and low output resistance. Because of the very high differential gain (100 000), only a tiny difference in levels will exist between the two input pins of the op-amp.

Suppose the output level is +2 V, then from the basic formula:

$$V_o = A_v (V_1 - V_2)$$

we get $(V_1 - V_2) = V_o/A_v = \dfrac{2}{100\,000} = 0.02\,\text{mV}$

Therefore the difference between V_1, the signal input, and V_o (the same connection as V_2) is very small. In other words, the output "follows" the input. The high input resistance also results because of the feedback from output to inverting input. R_{in} is very high for the reason that only a tiny input current is necessary to provide the small "difference" voltage between the op-amp input pins to set up the required output. This high resistance input means that there will be hardly any distortion of the signal from the sensor. In addition, the low output resistance of the op-amp is able to supply the current required by R_L.

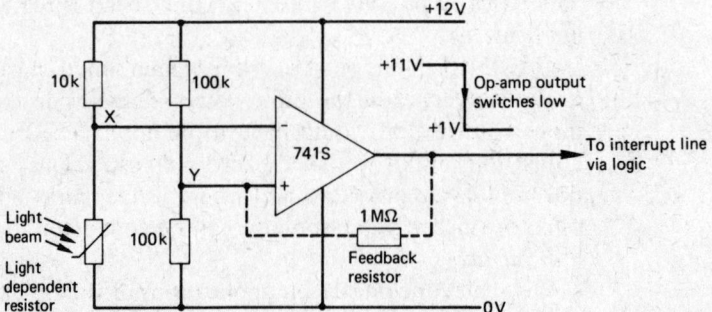

Fig. 2.14b Op-amp as comparator. When the light beam is broken, the photocell resistance rises and the voltage at X becomes greater than Y. The op-amp output switches low.

Another common application of op-amps is as a **comparator** (see *fig. 2.14b*). In this example the non-inverting terminal is held at a fixed voltage (+6 V) and, while the light beam is falling onto the ORP12 (a light-dependent resistor), the voltage on the inverting input will be lower, at say 1 V. The op-amp output will therefore be driven into positive saturation and will be at about +11 V. If the light beam is broken, the LDR resistance rises, and the voltage at point X will then exceed that at point Y. This will cause the op-amp output to switch rapidly to its negative saturation level at about +1 V. This change of state can be used, after shaping by a logic gate, as the interrupt for a micro system. There are, of course, several other possibilities with this circuit since the sensor could be a thermistor to give an output change of state when a set value of temperature is exceeded, or a smoke/gas sensor with the output driving an alarm.

One of the desirable features of a comparator circuit is high speed, and this means that an op-amp with fast slew-rate is required. The 531 (slew-rate of 35 V/μsec) or the 741S (slew-rate 20 V/μsec) would be suitable choices; but if really high speed is required a specially designed comparator IC such as a 311 or 710 would have to be used. Another feature of the circuit is the inclusion of a small amount of hysteresis set up by the 1 MΩ feedback resistor. This should ensure jitter-free triggering when the input level has just crossed the trip point, because this feedback resistor causes a small drop in the reference level at point Y when the output switches.

Fig. 2.14c Amplifier circuit: non-inverting

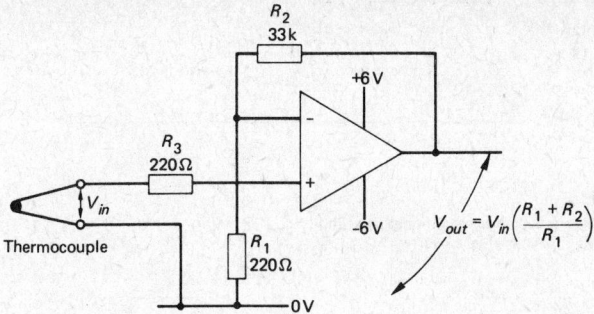

The circuit in *fig. 2.14c* shows an op-amp used as a **linear non-inverting amplifier** to increase the small signal obtained from a thermocouple. The thermocouple gives an output voltage of about $40\,\mu V/°C$ which, with the values given, would be increased by a factor of 150. In the non-inverting amplifier, a portion of the output signal is fed back to the inverting terminal to oppose the input. This negative feedback arrangement gives a stable fixed value of voltage gain that depends on the values of R_1 and R_2:

$$\text{Voltage gain } A_v = \frac{R_1 + R_2}{R_1}$$

Fig. 2.14d Inverting amplifier

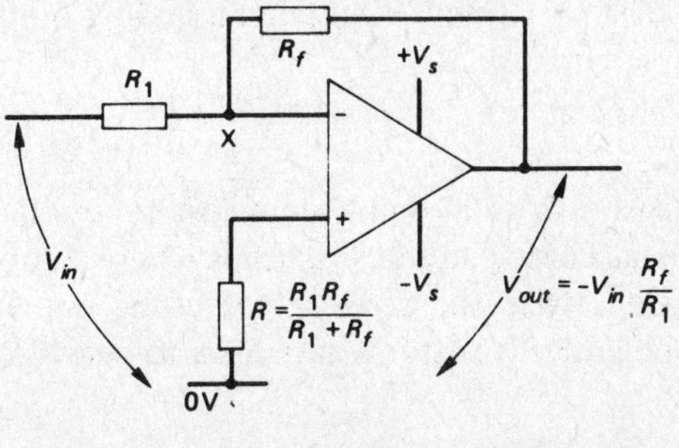

The **inverting amplifier** configuration (*fig. 2.14d*), the opposite to the previous example, again has the negative feedback connected from output to the inverting terminal but the input is applied via a resistor R_1 to the same point. Because of the very high internal gain of the op-amp, the common point of R_1 and R_f hardly changes at all, even when the output is a few volts. For this reason point X is often referred to as a "virtual earth" point. The gain of the arrangement is fixed by the resistor values to R_f/R_1.

The circuit can be extended to make a summing amplifier, and this in turn can be used as the basis of a simple **4-bit digital-to-analog convertor**. Each of the input resistors is weighted in value in a binary sequence, i.e. R, 2R, 4R and 8R, and the digital word to be converted is used to operate electronic switches which connect either a fixed reference voltage $(+5\,V)$ or zero volts to the input resistors. For example, suppose the digital word to be converted is

Fig. 2.14e Inverting summer

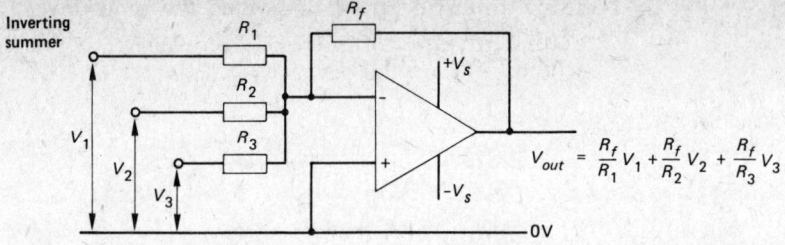

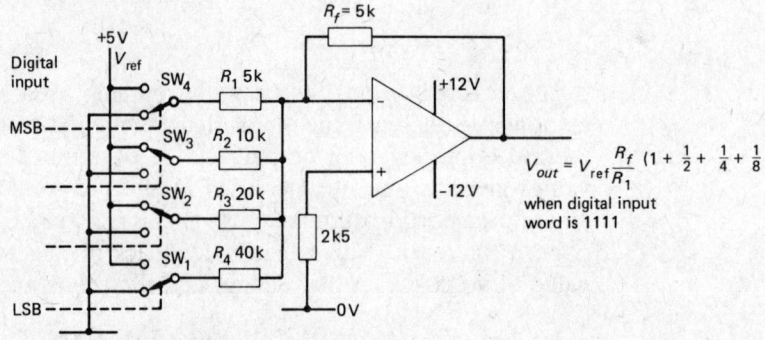

0110. Switches 2 and 3 (fig. 2.14e) will operate to connect the reference voltage to resistors R_2 and R_3 giving an analog output voltage of

$$V_o = \frac{R_f}{R_1} V_{ref} (0 + \tfrac{1}{2} + \tfrac{1}{4} + 0) = 3.75\,\text{V}$$

The DAC shown is a simple low-cost type which can only be used for relatively low accuracy and in situations where there are only a few bits to be converted. Otherwise the resistor values become unwieldy (to convert an 8-bit word the largest resistor would have to be 640 kΩ) and fairly large errors can occur.

2.4 Power-switching Devices

Many control systems have the requirement of an interface capable of switching relatively large values of current and at high voltages. Power switching devices can be grouped as

Power transistors (bipolar)
Darlington connected transistors
Power FETs (VMOS, HEXFET and TMOS)
Thyristors and triacs.

One important point is that a power switching device will probably have to be mounted onto a suitable heat sink. This is a piece of metal which is a good heat conductor (usually of black anodised aluminium) and which enables the heat generated at the switch junction by the high current to be efficiently

Fig. 2.15 Heat sinks

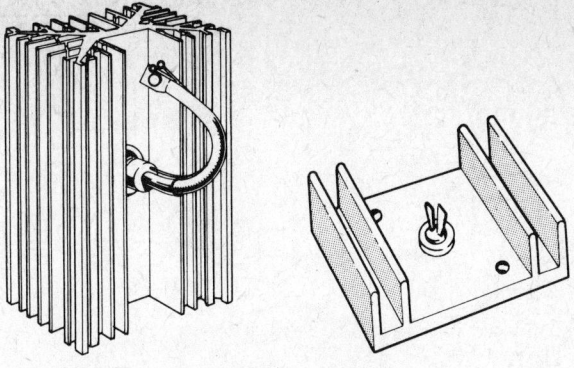

Fig. 2.16 Power transistor outlines

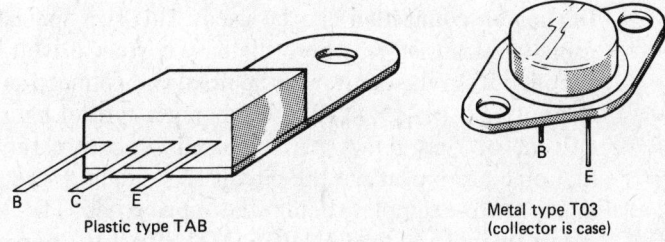

Plastic type TAB

Metal type T03 (collector is case)

conducted away (see *fig. 2.15*). In this way a dangerous rise in the temperature of the switching device is avoided.

A power transistor operates in exactly the same fashion as a small signal type but is constructed so that it can handle the power dissipation required. A relatively large volume of silicon is used and the collector will be connected to the metal can or the tab. Typical outlines are shown in *fig. 2.16*. The data for some common types is as follows:

Type	P_T (max. power)	I_C (max. current)	V_{CEO} (max. volts)	h_{FE} (min. current gain)
TIP 33/34 npn/pnp	80 W	10 A	60 V	10
BD 131/132 npn/pnp	15 W	3 A	45 V	20
2N3055 } pnp 3055 }	115 W	15 A	60 V	20

Note that the maximum power rating is quoted at the temperature of the transistor's case. Another point is that the current gain of a power transistor tends to be low. This means that the drive current required at the base can be fairly high—100 mA to control 1 A in the collector if the current gain is only 10.

Fig. 2.17 The Darlington circuit

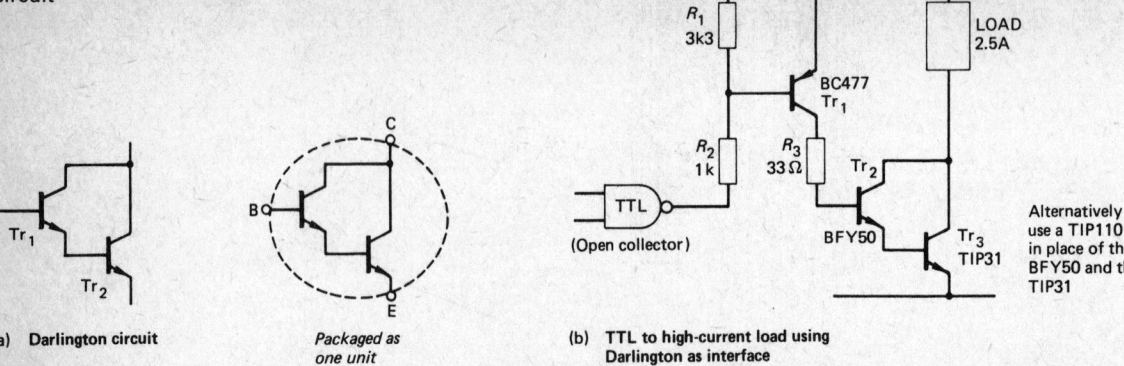

(a) Darlington circuit

Packaged as one unit

(b) TTL to high-current load using Darlington as interface

In situations where the available drive current is limited, then the **Darlington connection** can be used. This is a special circuit (see *fig. 2.17a*) using two transistors, where the base current drive to the output transistor is supplied from the emitter of the first. The connection has a very high value of current gain ($h_{FE1} \times h_{FE2}$), 1000 being a typical figure. In this way an input current of only 1 mA can be used to control an output of 1 A. Many manufacturers package the circuit as a single discrete component with three leads, two examples being the npn types TIP110 (TIP115 is the pnp complement) and the MJ11015 (MJ11016 is the pnp complement). These can dissipate 50 W and 200 W and have minimum current gains of 500 and 1000 respectively.

An interface example from **TTL logic**, to a 2.5 A load, is shown in *fig. 2.17b*. When the TTL gate output is high, there is no base current to Tr_1 and this transistor is off. The Darlington formed by Tr_2 and Tr_3 will also be off since the base current drive to Tr_2 is the collector current of Tr_1. When the TTL gate switches to a low output (logic $\emptyset$), current is passed from the +5 V rail through the base/emitter junction of Tr_1 and R_2. An amplified version of this current flows out of Tr_1 collector into the Darlington, forcing both Tr_2 and Tr_3 to switch on. To set up the required load current of 2.5 A, the emitter current of Tr_2 is about 250 mA and the collector current of Tr_1 is 8 mA. Consequently the TTL gate has to sink only about 2 mA to ensure that the load is fully on.

Mention has already been made of the potential of the enhancement mode **power FET** in interfacing. The structure of one type, the vertical FET (VFET), is shown in *fig. 2.18*. The cross-section shows that the geometry is similar to a planar transistor except that a V-groove is etched through the top n and p regions. On top of this V-groove is a coating of silicon dioxide which acts as the insulator between the metal of the gate and the body of the device. With the drain positive with respect to the source, a positive voltage on the gate will induce the regions in the p material opposite the groove (both sides) to invert to n-type. A large current will then flow from drain to source. The short thick channels created enable the VFET to pass the high values of drain current and this is controlled by the input voltage between gate and source.

Fig. 2.18 Structure of the VFET

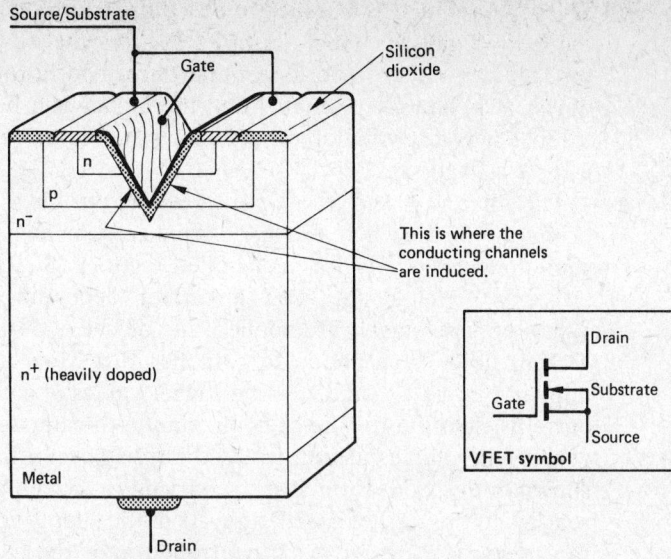

The parameter of interest is called g_{fs} (the forward conductance):

$$g_{fs} = \frac{\Delta I_d}{\Delta V_{GS}} \text{ milliSiemens} \qquad \text{(old units were mA/V)}$$

A typical value is 200 mS.

Using power FETs is very straightforward. An example of an interface from CMOS logic to drive a 3 A d.c. motor is shown in *fig. 2.19*. Since the FETs are voltage controlled, they present only a light load to the CMOS output. The 220 Ω resistors are included to prevent any unwanted high-frequency oscillations between the parallel-connected FETs. This connection method can be used because the power FETs will tend to share the load equally, which is not something which the bipolar transistor is good at doing.

Fig. 2.19 Example of the use of power FET

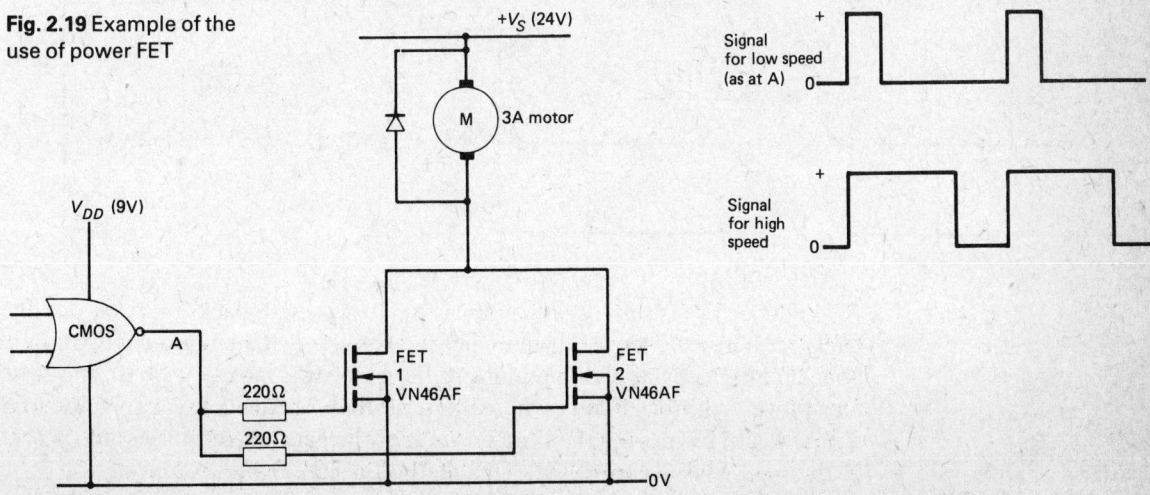

The motor control is assumed to be a pulse width type. When the output from the CMOS gate goes high, both FETs are switched on and current is passed through the motor. The longer the on period in one cycle, the greater the power switched to the motor and the higher will be its speed. If the signal frequency is a few hundred hertz, then the motor will average out the pulses to give a fixed speed.

The thyristor and triac are power-controlling semiconductor devices, primarily for use where the load is connected in an a.c. supply. The **thyristor**, sometimes called a silicon controlled rectifier (SCR), can be made to act as either an "open circuit" or as a rectifier, depending on how its gate is used. When no gate signals are applied, the device is off and only a small leakage current flows from anode to cathode. But when a pulse of gate power is applied (above V_{GT} and I_{GT}, the trigger values of gate to cathode voltage and current) then the resistance of the anode-to-cathode path falls to a very low value, and a large current, set by the resistance of the load, will flow. This is shown in *fig. 2.20*. Note that, by varying the time position of the gate pulses relative to the start of each half-cycle of the supply voltage, the point at which the thyristor "fires" can be controlled and, therefore, also the amount of power dissipated in the load. The thyristor latches on once switched, but will turn off as the rectified a.c. waveform returns to zero.

Fig. 2.20 Thyristor action

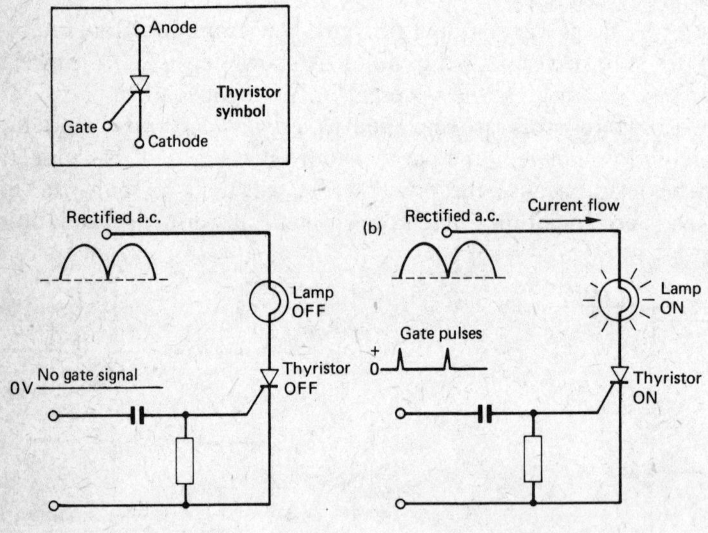

Fig. 2.21 The triac

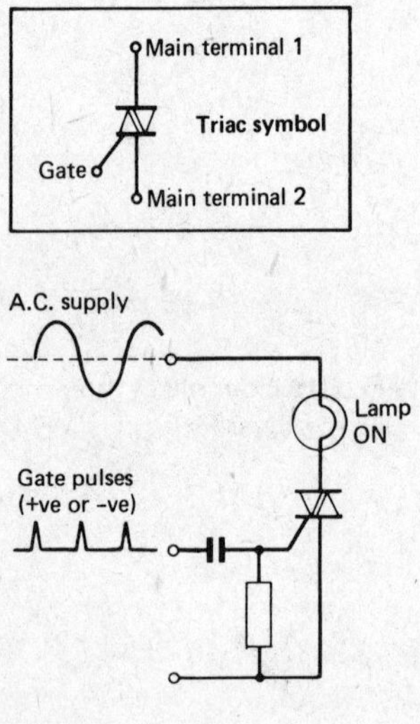

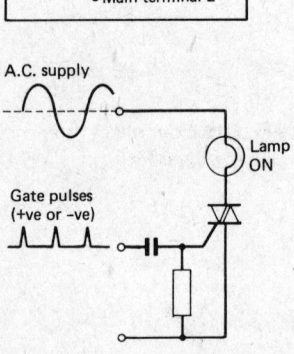

The **triac** is essentially two thyristors connected back to back in one package. This very useful device can therefore be switched into conduction in *both* directions by applying a pulse to the gate. It is mostly used in full-wave a.c. power control circuits where there is then no need for a rectifier (*fig. 2.21*). Again by moving the time position of the gate trigger pulses, the power in the load can be controlled. This method is called *phase control*.

2.5 Useful Analog Circuits

Any electronic system will be built up from several smaller sub-units, each of which can usually be given an identity tag—there will be names such as clock pulse generator; signal shaper; filter; and so on. It would not be possible to list all the various circuits in use in this one section, but some are so standard that a brief look at them is warranted.

1 First of all, consider the circuits that can be used for **waveform and signal generation**:

Sine wave oscillators
Crystal controlled oscillators
Square wave and clock pulse generators
Triangle and ramp generators.

A basic oscillator circuit will consist of an amplifying device with a frequency-determining network and positive feedback (*fig. 2.22*). The frequency-determining components are either a tuned circuit (an inductor and capacitor in parallel) or a *CR* phase shifting network. The **Wien bridge oscillator** is a good example of the latter type and is an excellent circuit for

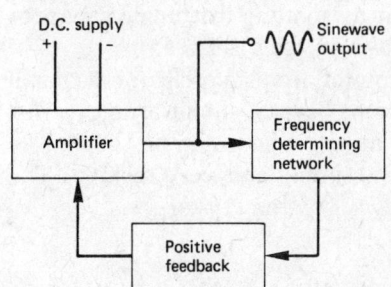

Fig. 2.22 Block diagram of basic oscillator

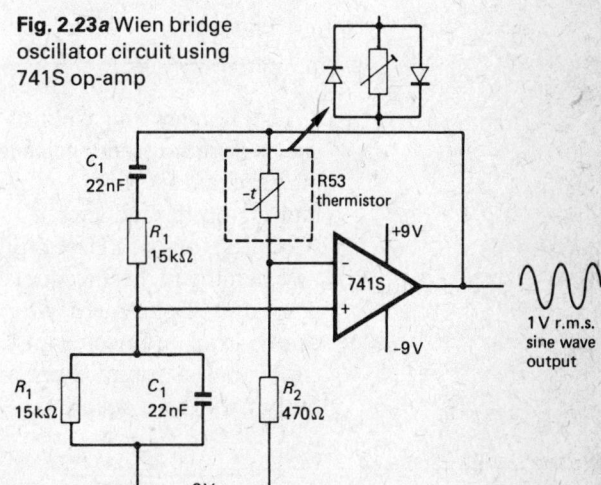

Fig. 2.23a Wien bridge oscillator circuit using 741S op-amp

generating medium-frequency (0.1 Hz to 1 MHz) low-distortion sine waves. A version using a 741S op-amp is shown in *fig. 2.23a*. The frequency-determining network is formed by the series *CR* and parallel *CR* combination. These give zero phase shift at one particular frequency from the output to the non-inverting input and thus supply the necessary positive feedback:

$$f_0 = \frac{1}{2\pi CR}$$

With the values given, the output will oscillate at approximately 1 kHz. Different values of capacitor and/or resistor can be used to vary this frequency. These components could be switched into circuit using a digital output from a micro system. A stabilising component, the enclosed thermistor type RA53, is used to set the amplifier's overall gain to just make up for the losses in the frequency-determining network. The output will be a 1 V r.m.s. sine wave with quite low distortion. If a pure sine wave is not required then the thermistor (which is expensive) can be replaced by a diode shaping network as shown in the insert.

Fig. 2.23b Colpitts oscillator circuit

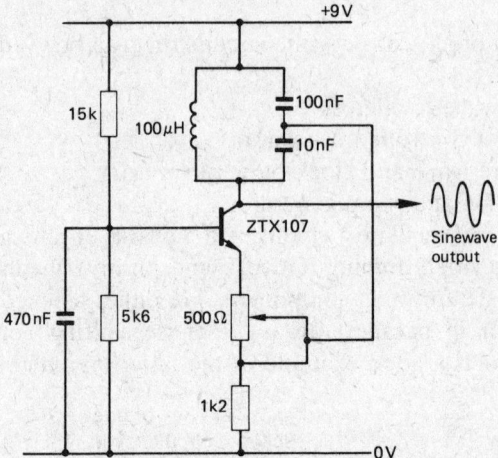

The **Hartley** and **Colpitts** are typical of oscillators used to generate sine waves at radio frequencies and these use an *LC* resonant tuned circuit as the load for an amplifier. A portion of the oscillating signal is fed back via a tapping on the inductor (for the Hartley) or by splitting the tuning capacitor (Colpitts). A 175 kHz Colpitts circuit is shown in *fig. 2.23b*.

When the highest frequency stability is required from an oscillator, a crystal is used as the frequency-determining network. Several circuit arrangements can be used and two examples of such oscillators are shown in *fig. 2.23c*. The circuit based round the CMOS 4001B IC is simple and very reliable. The output will be a square wave at a frequency set by the crystal.

Fig. 2.23c High frequency stability circuits using a crystal

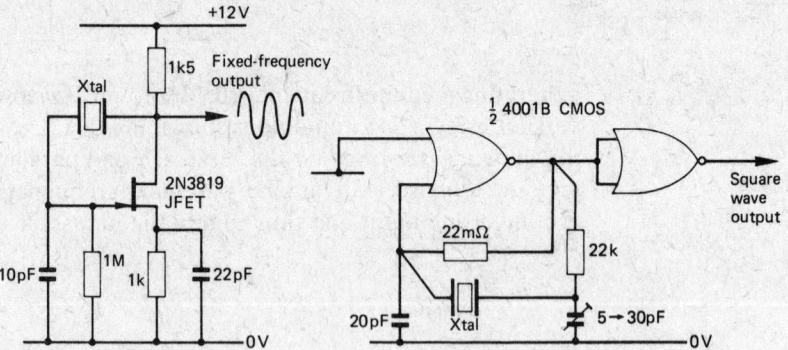

2 Signal conditioning is another of the important requirements in most systems. This can take the forms of reshaping logic pulses which have been degraded during transmission, of changing the levels of a signal, or of removing superimposed noise.

Fig. 2.24 Schmitt trigger circuit

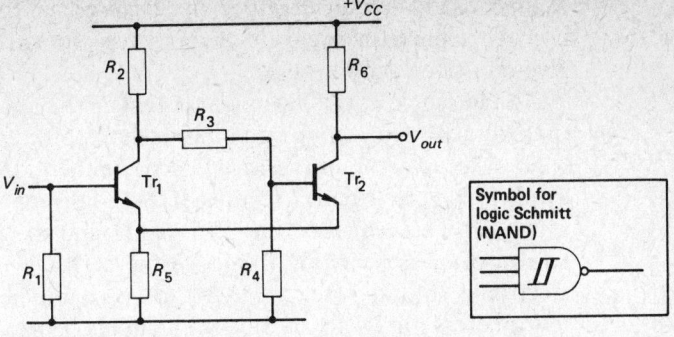

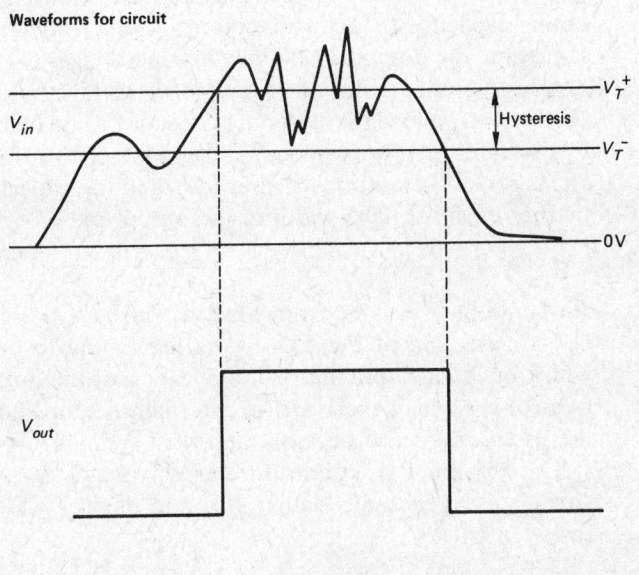

The **Schmitt trigger** is the standard circuit for performing these tasks and it is such a useful circuit that a version of it is included in most logic families (for example the TTL 7413 and CMOS 4093). The circuit (*fig. 2.24*) is formed around two transistors which have their emitters joined together, giving positive feedback. This arrangement gives a snap-action switch with Tr_2 on while the input is at a lower level than the positive going threshold (V_T+), and with Tr_1 off. The value of V_T+ is set by the potential divider R_2, R_3 and R_4. As soon as the input level exceeds V_T+, Tr_1 switches on and turns Tr_2 off, causing the output to rise smartly to the positive rail level. The circuit remains in this state until the input falls below a level called the negative going threshold V_T-. The difference between the two threshold levels gives the circuit a useful feature of hysteresis. This is illustrated in the waveform diagrams where the noise present on the input signal is ignored because its excursions are contained within the hysteresis of the Schmitt. This also shows how the Schmitt sharpens up the poor rise and fall times of the input signal to give one clean output pulse.

Another circuit, called the **monostable**, will give one output pulse of fixed time duration when triggered on by a short spike at its input. The pulse width is set by a capacitor and resistor which can therefore be programmed. As well as the discrete circuit, several IC versions of the monostable are also available. These include the TTL 74121 and the CMOS 4047, but the best known of all is probably the 555 Timer. This IC contains all the necessary electronic "blocks" (see *fig. 2.25*) to give quite accurate time delays from a few microseconds up to several minutes, controlled by an external CR network. Without a trigger pulse applied, the internal bistable is reset with $\overline{Q} = 1$, and this forces the internal discharge transistor to be on. The external timing capacitor is then clamped near zero volts and cannot charge. At the same time the output is held low. When a trigger pulse is applied, the internal bistable memory is set, the output switches high, and the discharge transistor switches off, allowing the timing capacitor to be charged. The voltage across C_t rises until it just exceeds $\tfrac{2}{3} V_{CC}$, at which point the internal bistable of the 555 is reset, the output switches low, and the capacitor is rapidly discharged by the transistor. The width of the output pulse is given by the formula:

$$T \simeq 1.1\, C_t R_t$$

R_t should have a value from $1\,\text{k}\Omega$ min up to about $1\,\text{M}\Omega$.

The reset pin of the 555 allows the timing to be externally terminated, whilst the control pin allows modification of the output width by application of a voltage which overrides the internally set threshold level. Both these pins are useful in microprocessor-controlled systems.

The 555 timer IC can also be wired up as a free-running square wave or clock pulse generator. Other ICs and devices which can be used for this include:

Transistor astable multivibrator
Unijunction relaxation oscillator
Schmitt invertor or NAND gate with positive feedback
CMOS NAND and NOR gates
Op-amp wired with positive feedback
Specialised ICs such as the CMOS 4047.

Useful Electronic Devices and Circuits 49

Fig. 2.25 The 555 timer IC

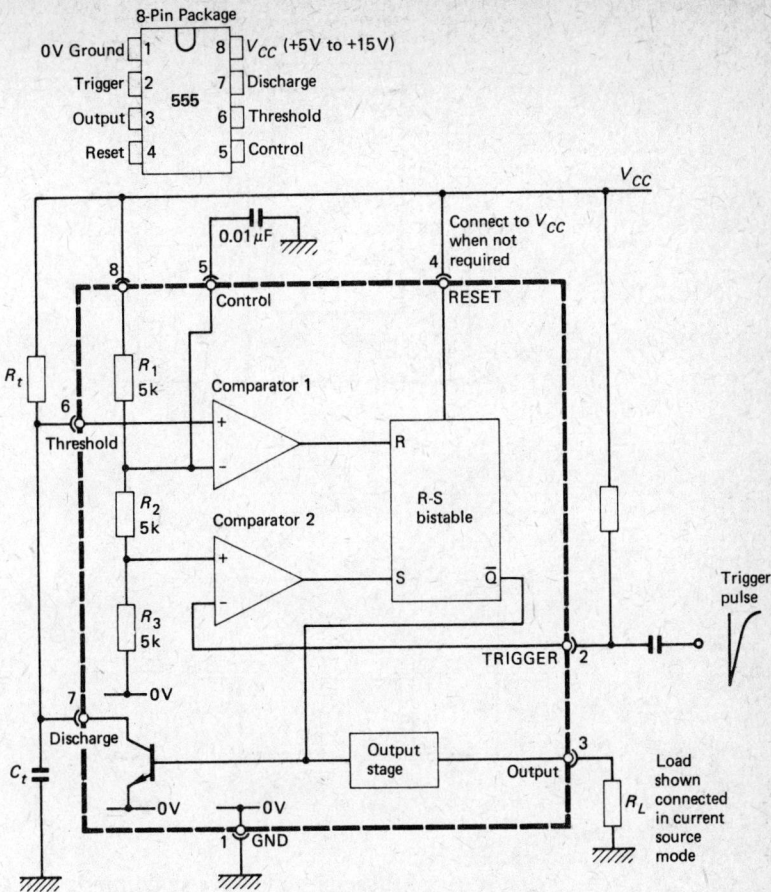

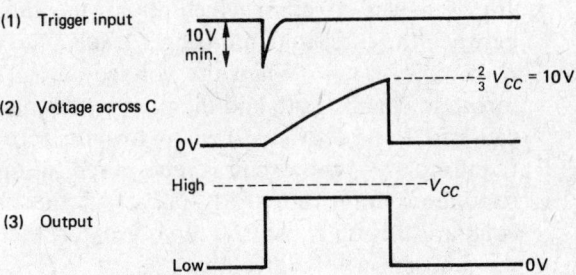

Fig. 2.26 The 555 wired in astable mode

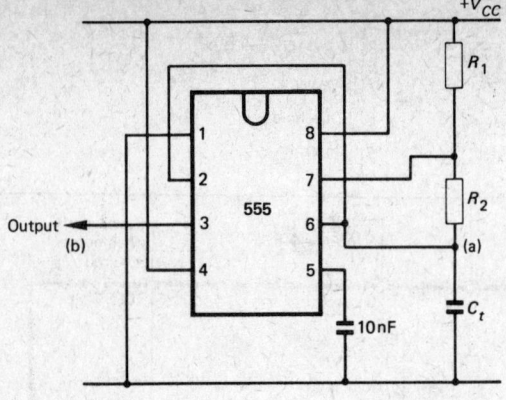

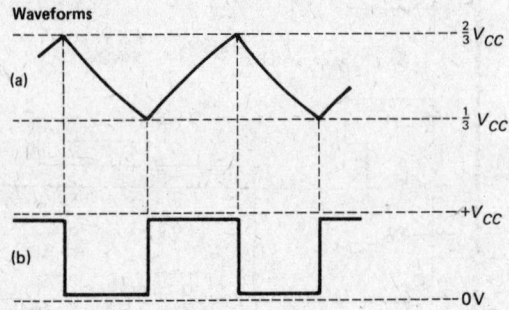

What most of these circuits have in common is a *CR* timing network which will determine the frequency and probably also the mark-to-space ratio of the output waveform. Take the 555 wired-in **astable** mode (*fig. 2.26*). Pins 2 and 6 are connected together which allows the capacitor to charge and discharge between the threshold and trigger levels. At switch-on, C_t charges via R_1 and R_2 towards $+V_{CC}$. When the voltage across C_t reaches $\frac{2}{3}V_{CC}$, the output is forced to change state and the internal discharge transistor is turned on. This causes C_t to be discharged via R_2 towards zero volts. When the voltage across C_t falls to $\frac{1}{3}V_{CC}$, the circuit is retriggered to start the cycle again. In this way a continuous train of pulses appears at the output. By making R_2 large in comparison with R_1, say 50 times greater, the frequency of the nearly-square wave output will be given by

$$f \simeq \frac{1}{1.4\, R_2 C_t}$$

Fig. 2.27 Examples of clock pulse generators

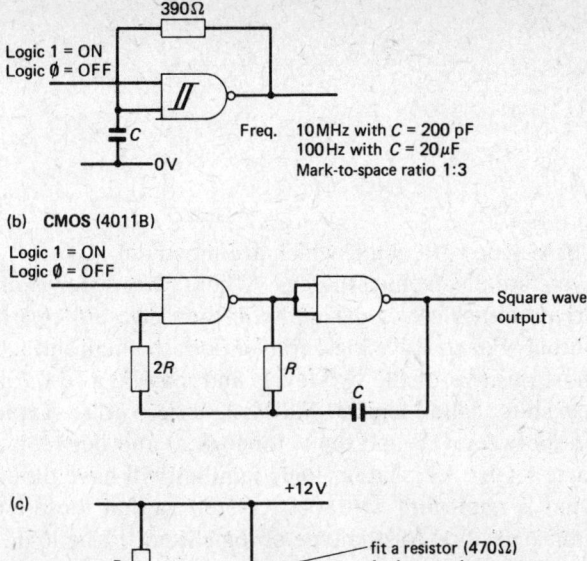

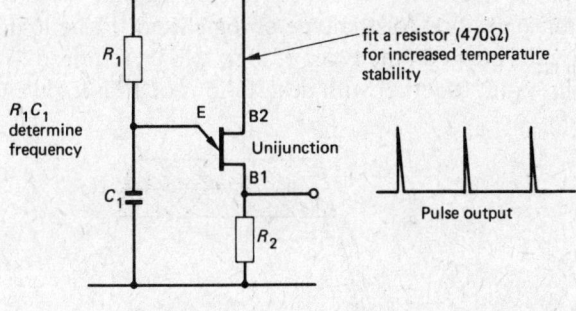

Other examples of clock pulse circuits are shown in *fig. 2.27*, most of which can be gated on and off by a controlling logic signal—very useful in control systems applications.

One final point; this chapter was intended only as an introduction to a few of the devices and circuits in common use. Other examples will be covered in later chapters.

3 Digital Circuits

3.1 Introduction

These types of circuit, which are mostly in IC form, are built up using many two-state electronic switches. When a transistor is used as a switch it can be driven either ON or OFF and will then give either a low-state or a high-state output. For positive logic convention, the high state, called **logic 1**, will be the most positive of the two levels and may be a few volts positive; whereas the low state, called **logic 0**, will be near zero volts. Typical values for TTL gates are between 0 V and 0.4 V for logic 0 and between +2.4 V and +3.3 V for logic 1 (*fig.* 3.1). Other logic families will have different levels from this but what is important within any system is that these levels are held within the limits specified for the type of logic used. If the logic is mixed, then interface circuits between the types of logic will be required to translate the logic levels. This point together with descriptions of TTL logic and CMOS logic is covered later.

Fig. 3.1 Positive logic

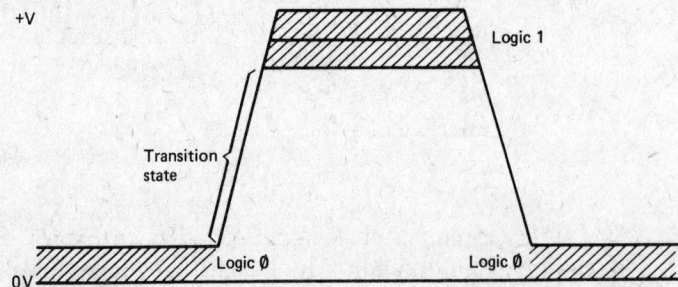

Some of the important advantages of digital systems in comparison with analog are:

a) A switch is either ON or OFF and should not give a level in between. This reduces the uncertainty about the presence of a signal.
b) For the same reason, digital signals are less susceptible to noise and interference.
c) Data in digital form (groups of 1s and 0s) can be more easily stored, transmitted, processed and reshaped than analog.

There are two main forms of digital logic:

1 COMBINATIONAL LOGIC in which a set of input conditions must be simultaneously combined in order to give a particular output. For example, to

Fig. 3.2 Alarm circuit using combinational logic

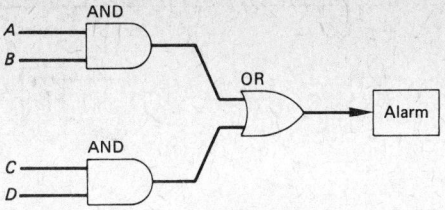

The Alarm will operate if signals (*A* and *B*) OR signals (*C* and *D*) are present

get an alarm to operate, the inputs to the logic might have to be (*A* and *B*) or (*C* and *D*). This function requires two AND gates and one OR gate as shown in Fig. 3.2.

2 SEQUENTIAL LOGIC in which logic circuits have a built-in memory and give an output, in response to an input, which is dependent on the circuit's previous state. Typical examples of sequential systems are counters and shift registers, and the basic building block is the bistable.

In general, a complete system will be a mix of combinational and sequential logic, all of which will be built up from the basic gates.

3.2 Logic Gates

A **gate** in digital terms is a circuit which gives a certain logic output as long as a particular set or a combination of states exists on its inputs. The number of inputs can vary from one up to 16 or higher but for convenience we shall assume 2-input elements.

The types of logic gate are:

AND, OR, NOT, NAND, NOR, and EXCLUSIVE-OR.

The logic symbols for these gates are shown in *fig. 3.3*.

The **AND gate** gives a logic 1 output only when all its inputs are at logic 1. In other words, for a two-input gate the output will be high only when input *A* AND input *B* are high. The output will be logic ∅ and low for all other input conditions.

The **OR gate**, sometimes referred to as the inclusive-OR, will give a logic 1 output if there is a logic 1 on any of its inputs. For a two-input OR gate, the ouput is high if input *A* OR input *B* is high.

The **NOT gate** or invertor is a single-input gate that always give an ouput which is the opposite to its input. If its input is logic 1, then its output will be logic ∅, and vice versa.

Following the AND gate with a NOT gate produces the NAND function. A **NAND gate** will give logic ∅ only when all of its inputs are high at logic 1. For the two-input NAND this means that both *A* and *B* must be 1 to give a logic ∅ output.

In a similar way, the **NOR gate** is created by following an OR gate with a NOT gate. A NOR gate will give a logic ∅ output if any input is a logic 1.

The **exclusive-OR gate** is a special circuit which only gives a logic 1 output if either of its inputs is at logic 1 but not when they are both at 1 or both at ∅.

Fig. 3.3 Commonly used symbols for logic gates

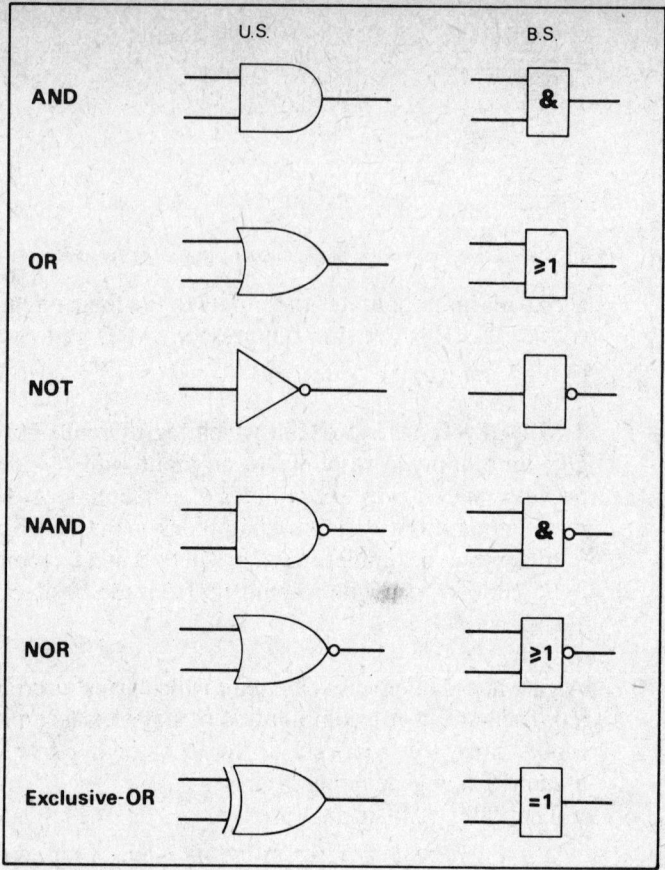

Fig. 3.4 Truth table for logic gates (only two-input gates assumed)

Inputs A	B	AND	OR	NAND	NOR	Exclusive-OR
∅	∅	∅	∅	1	1	∅
∅	1	∅	1	1	∅	1
1	∅	∅	1	1	∅	1
1	1	1	1	∅	∅	∅

The operation of the logic gates can also be followed from a truth table. This is a table that shows the resulting output for each and every possible combination of inputs. It can be a useful aid in following through the operation of a combinational logic circuit. The truth table for all the gates is shown in *fig. 3.4*.

3.3 Boolean Algebra

Invented by George Boole in the 19th century this type of algebra is very useful in describing, analysing and simplifying logic. It is based on logical statements which are either true or false, and can therefore be used as a mathematical tool in the design and analysis of the two-state logic.

The notation is very simple. In Boolean algebra, the statement "operate switches A and B and C to get an output F" is written

$$F = A.B.C$$

The AND function is represented by the dot symbol (.).

Similarly the statement "operate either switch C or D to get an output F" is written

$$F = C + D$$

The OR function is represented by the plus symbol (+).

The NOT function is represented by a bar over the switch:

$F = \overline{A}$ means "do *not* operate switch A to obtain an output"

Using this

the NOR function is $F = \overline{A+B}$ (NOT-OR)

the NAND function is $F = \overline{A.B}$ (NOT-AND)

Now consider some examples of Boolean logic statements.

a) $F = A.B + \overline{C}$ means "operate A and B together, or not C, to get an ouput at F."

b) $F = A.(B + \overline{C})$ means "operate A and B, or A and not C, to get an ouput at F."

You can see that it is much easier to write the circuit function down in Boolean Algebra rather than in words. The logic diagrams for the two examples are shown in *fig. 3.5*.

Fig. 3.5 Circuits illustrating logic functions

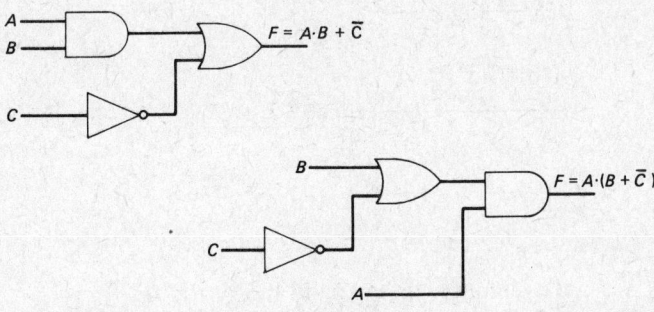

Boolean Algebra has a set of self-evident truths, identities and rules which are listed in *fig. 3.6* and we shall use these in a few simplification examples.

Fig. 3.6 Identities and rules of Boolean Algebra

IDENTITIES

$A + 0 = A$ $A.0 = 0$
$A + A = A$ $A.1 = A$
$A + 1 = 1$ $A.A = A$
$A + \bar{A} = 1$ $A.\bar{A} = 0$

RULES

$A + B = B + A$ $(A+B) + C = A + (B+C)$
$A.B = B.A$ $(A.B).C = A.(B.C)$

$A + A.B = A$ since $A.(1+B) = A.1$
$A.(A+B) = A$ since $A.A + A.B = A + A.B$
$A + \bar{A}.B = A + B$

$\overline{A+B} = \bar{A}.\bar{B}$ de Morgan's rules

$\overline{A.B} = \bar{A} + \bar{B}$

Example 1. $F = (A+B).(A+C).\bar{B}$

The logic diagram for this, together with its simplified form is shown in *fig. 3.7*.

Fig. 3.7

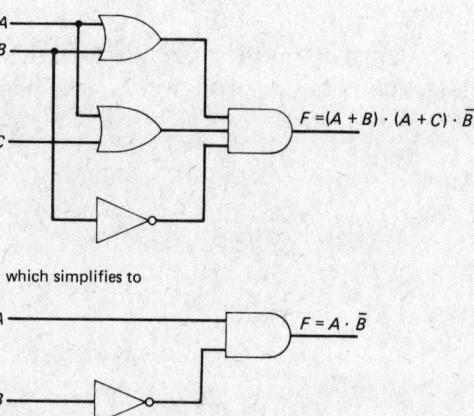

which simplifies to

To simplify, first expand the brackets:

$F = (A.A + A.C + B.A + B.C).\bar{B}$

Next step: multiply variables inside the bracket by $\bar{B}$:

$F = A.A.\bar{B} + A.C.\bar{B} + B.A.\bar{B} + B.C.\bar{B}$

Next step: since $B.\overline{B} = \emptyset$ (open circuit condition)
and $A.\emptyset = \emptyset$
also $C.\emptyset = \emptyset$

then $F = A.\overline{B} + A.C.\overline{B}$

Next step: factorise since $A\overline{B}$ is common:

$$F = A.\overline{B}.(1+C)$$

But $1+C = 1$ and $A.\overline{B}.1 = A.\overline{B}$

$$\therefore F = A.\overline{B}$$

Fig. 3.8

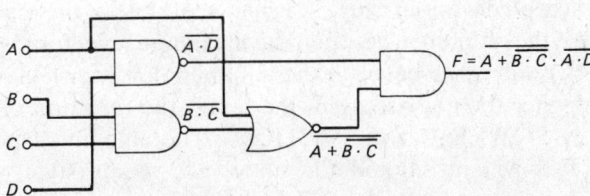

which simplifies to

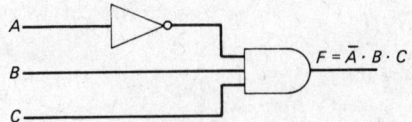

Example 2. $F = \overline{A + B.C}.(\overline{A.D})$ (unsimplified diagram fig. 3.8)

To simplify we must use de Morgan's theorems.
Step 1: remove bars:

$$F = \overline{A}.\overline{\overline{B.C}}.(\overline{A} + \overline{D})$$

Next step: $\overline{\overline{B.C}} = B.C$.

$$\therefore F = \overline{A}.B.C.(\overline{A} + \overline{D})$$

Next step: remove bracket:

$$F = \overline{A}.B.C.\overline{A} + \overline{A}.B.C.\overline{D}$$
$$= \overline{A}.B.C + \overline{A}.B.C.D$$

Next step: factorise since $\overline{A}.B.C$ is common:

$$F = \overline{A}.B.C.(1 + D)$$

But $1 + D = 1$

$$\therefore F = \overline{A}.B.C$$

58 Practical Interface Circuits for Micros

3.4 Logic Conventions and Parameters

We have already seen that, if *positive logic* convention is used, logic 1 will be the most positive voltage and logic Ø the most negative voltage level. These levels will be different depending on the type of logic family being used, whether TTL, CMOS or ECL. In addition there are other properties and parameters which are used to specify logic performance. These are illustrated in *fig. 3.9*.

1. **FAN-IN** The number of inputs that can be accommodated on a gate without the logic levels going out of specification.
2. **FAN-OUT** The ability of a gate to drive several inputs of similar gates simultaneously without the logic level at the ouput of that gate going out of the specified limits.
3. **PROPAGATION DELAY TIME** The speed at which the logic switches.
4. **NOISE MARGIN** The measure (in volts) of a noise signal that can be accepted without causing a change of state at the ouput. The noise margin will be the difference between the maximum level for logic Ø and the threshold; or the difference between the minimum level for logic 1 and the threshold. The higher the noise margin, the better the rejection of noise.
5. **POWER CONSUMPTION** The amount of power taken by one gate. This will be quoted for static, i.e. steady state, conditions, and also for dynamic (switched) conditions.

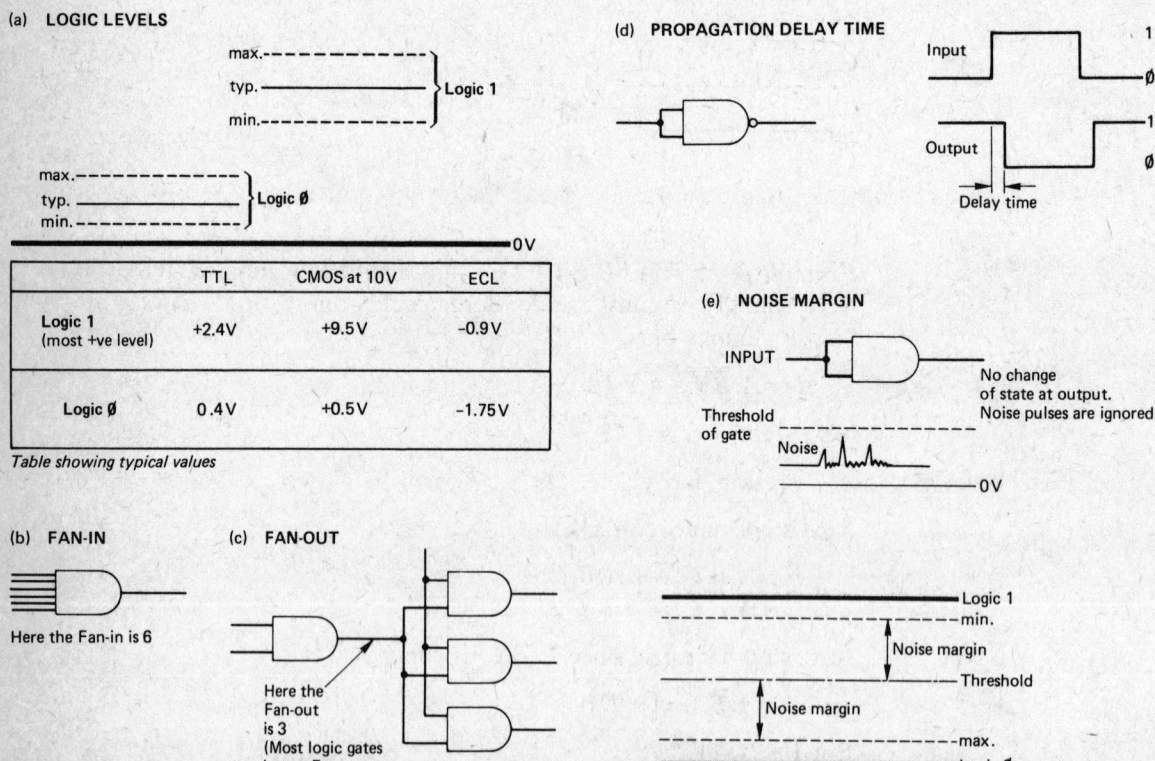

Fig. 3.9 Logic conventions and parameter

3.5 Logic Families

In order to make a complete working logic unit several digital ICs have to be linked together. It is then often more convenient to use, wherever possible, ICs of the same logic family; that is, either all TTL or all CMOS. Each of these types has its own particular advantages and main application areas. The trade-offs are usually those of switching speed versus power consumption, or switching speed versus noise margin. Generally, the faster the logic is required to operate, the greater will be its power consumption and the lower will be its immunity to noise.

1 TTL (Transistor Transistor Logic) is a popular and widely used logic family. It combines fast speed with moderate levels of power consumption and reasonable levels of noise margin. The main types are:

Standard TTL 74 series
Schottky TTL 74S
Low Power Schottky TTL 74LS
Advanced Low Power Schottky TTL 74ALS

One of the basic circuits is the NAND gate (*fig. 3.10*) and this will be used to describe the TTL operation.

The inputs are to a multi-emitter transistor Tr_1 which gives the AND function. Tr_2 is a phase splitter and Tr_3 and Tr_4 form an ouput stage (given the appropriate name of *totem pole*) which has a low output resistance in both logic states.

If all inputs are high, then Tr_1 passes current from its collector to turn on Tr_2. With Tr_2 on, Tr_3 is forced to conduct while Tr_4 is held off. The output is therefore low at logic $\emptyset$.

If any input is taken low, that particular base/emitter junction conducts,

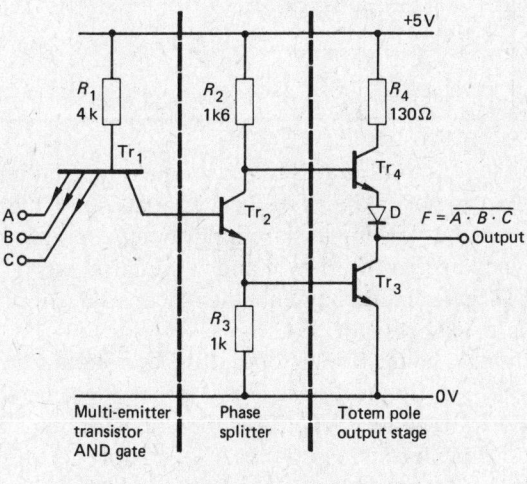

Fig. 3.10 Basic TTL NAND gate

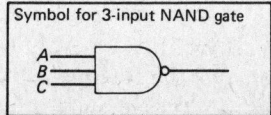

diverting current away from Tr_2. Both Tr_2 and Tr_3 turn off, while Tr_4 then conducts taking the output high to logic 1.

The action can be summarised as:
   all inputs high (1)   resulting output low (∅)
   any input(s) low (∅)   resulting ouput high (1)

The design for the circuit inside the IC gives TTL (standard type) the following typical performance figures:

Power consumption	10 mW
Noise margin	1 V
Propagation delay	10 nsec
Fan-out	10

The power supply to TTL should be a reasonably well regulated +5 V with an absolute maximum of 7 V. Any fault in the power supply which puts more than 7 V across a TTL IC will cause damage to the IC. For this reason, TTL power supplies are usually fitted with an overvoltage protection circuit.

As a TTL gate switches there is a brief instant when both output transistors (Tr_3 and Tr_4) conduct, causing a pulse of current to be drawn from the supply. "Spikes" on the supply rail can then be generated and may cause false triggering of other gates. However the problem can be easily overcome by decoupling the +5 V rail to 0 V using a 100 nF ceramic capacitor. In a TTL system several such decoupling capacitors will be wired in across the supply leads at suitable points, usually one capacitor for every four "gate" type TTL ICs (see *fig. 3.11*).

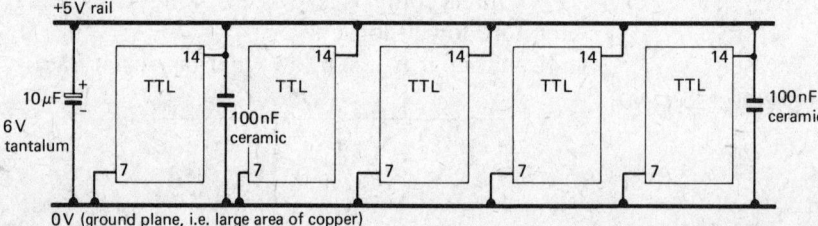

Fig. 3.11 Decoupling TTL chips

Any spare input, that is to say an unconnected input pin, on a TTL IC will assume a logic 1 state. This occurs because of the TTL multi-emitter input transistor; but leaving a spare input open circuit is not good practice because noise may be picked up by it and cause false outputs. Therefore, spare inputs to a TTL gate should be tied either to an adjacent driven input or to the +5 V rail via a 1 kΩ resistor.

Although being fast in operation the standard TTL circuit is not rapid enough for all applications. This is mainly because the internal transistors are allowed to saturate. When a transistor is driven hard on and saturated, its turn-off time is increased because it stores excess charge in its base. A Schottky transistor overcomes this problem by preventing the transistor from saturating. Schottky TTL operates in the same way as standard TTL but has a propagation delay of only 3 nsec (this type can be used for high-speed logic). Another very popular version, the low-power Schottky (74LS), switches in 7 nsec and only consumes 2 mW per gate. This can be used when supply power is limited.

2 **CMOS** logic (**Complementary Metal Oxide Silicon** field effect transistor logic) is the other versatile and popular logic family. It is made up of p and n channel enhancement mode mosfets and the type of construction results in a number of particular advantages when compared with TTL:

a) Very low power consumption—about 10 nW per gate for slow-speed operation.
b) A wide operating supply voltage range (from $+3\,V$ to $+18\,V$).
c) A very high fan-out.
d) Excellent noise immunity—the noise margin is typically 45% of the supply voltage.

CMOS is then ideal for use in battery-powered instruments, in electrically noisy environments, and where speed of operation is not the prime consideration. The disadvantages of CMOS compared with TTL are slower speed and the relatively high output resistance which does not enable it to drive external loads easily.

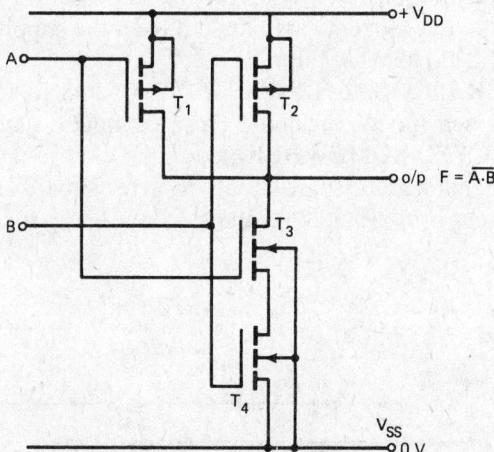

Fig. 3.12 CMOS NAND gate circuit

The operation of a CMOS NAND gate (*fig. 3.12*) is best described by a sort of truth table. First the definition for CMOS logic levels is:

Logic $\emptyset$ 0V to $0.3V_{DD}$ where V_{DD} is the supply voltage.
Logic 1 $0.7V_{DD}$ to $+V_{DD}$

With the NAND gate, the output can only be low when both T_3 and T_4 conduct and when both T_1 and T_2 are off. This condition only occurs when both inputs A and B are high at logic 1.

Inputs		State of MOSFETs				Resulting output
A	B	T_1	T_2	T_3	T_4	
$\emptyset$	$\emptyset$	ON	ON	OFF	OFF	1
$\emptyset$	1	ON	OFF	OFF	ON	1
1	$\emptyset$	OFF	ON	ON	OFF	1
1	1	OFF	OFF	ON	ON	$\emptyset$

CMOS ICs are operated by voltages (see MOSFET operation in Chapter 2) and therefore present only a very light load. This means that a very high fan-out can be achieved. If you study the table you will notice that for a logic 1 output either T_1 or T_2 is on at the same time that T_3 and T_4 are off; and that for a logic $\emptyset$ output both T_1 and T_2 are off; while T_3 and T_4 are on. Under static conditions there is no current path from the supply $+V_{DD}$ and 0 V. This is why the power consumption is low. Unfortunately, as the operating speed is increased, the power consumption rises because each time there is a switching change from $\emptyset$ to 1 and vice versa a small amount of power is taken from the supply (typically, power consumption is 1 mW/MHz per gate).

Because MOSFETs have only a thin insulating region between the gate and the body, the CMOS ICs can be easily damaged by electrostatic discharge. Although protection circuits are built in to prevent this sort of damage, care should always be taken in handling and soldering in CMOS. It is also very important that inputs to a gate or IC are not left open circuit. This is because a CMOS input is virtually a small capacitor in parallel with an extremely high resistance. If left open circuit, the capacitor charges up and puts the gate into its active region, in other words the internal MOSFETs all start to conduct. In this way a large current can be taken from the supply and the CMOS IC will overheat and probably burn out.

NEVER LEAVE AN INPUT TO A CMOS IC OPEN CIRCUIT.

All unused inputs, including those of unused gates in an IC, MUST BE CONNECTED SOMEWHERE.

As a general rule, spare inputs to gates should be connected to another driven input or to a suitable voltage level ($+V_{DD}$ for a NAND) and inputs to

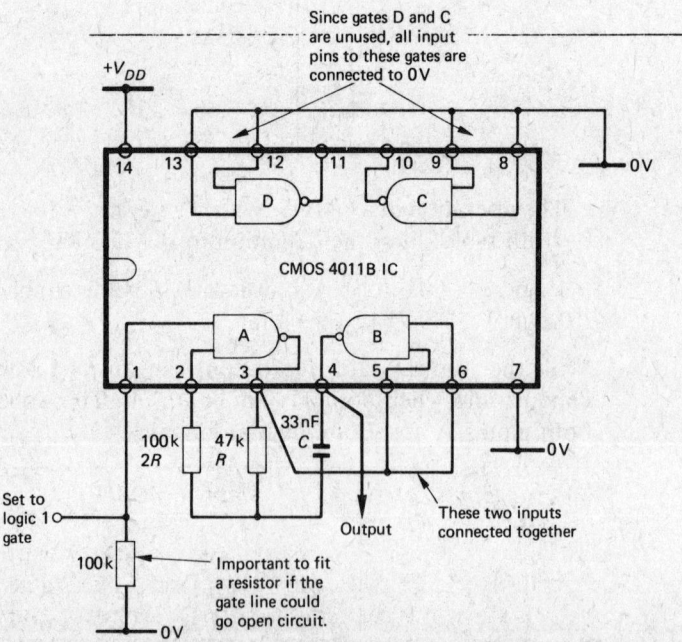

Fig. 3.13 Dealing with unused and spare inputs in a CMOS IC. In this example, only gates A and B are being used (as an oscillator)

unused gates should be disabled by connecting them to 0 V, or $+V_{DD}$ (fig. 3.13).

3.6 Interfacing Between Logic

There will obviously be situations where it is necessary to use a mix of TTL and CMOS logic ICs. Depending on the voltage levels, the following rules of thumb apply (fig. 3.14):

1 If the CMOS is running from a +5 V supply, then any TTL gate provided with a 2k2 pull-up resistor will drive an unlimited number of CMOS gate inputs.
2 Usually the CMOS will be running at a higher voltage, say +9 V, and then a TTL open collector output gate (7406 is an example) can be used with the pull-up resistor connected to the +9 V CMOS supply.
3 Any CMOS gate output, with a supply of +5 V, will drive *one* low power Schottky TTL input. Otherwise the CMOS (at a supply of +5 V or higher) must be interfaced to the TTL via a 4049B or 4050B buffer. These will drive any two TTL inputs.

Fig. 3.14 Interface methods from TTL to CMOS, and vice versa

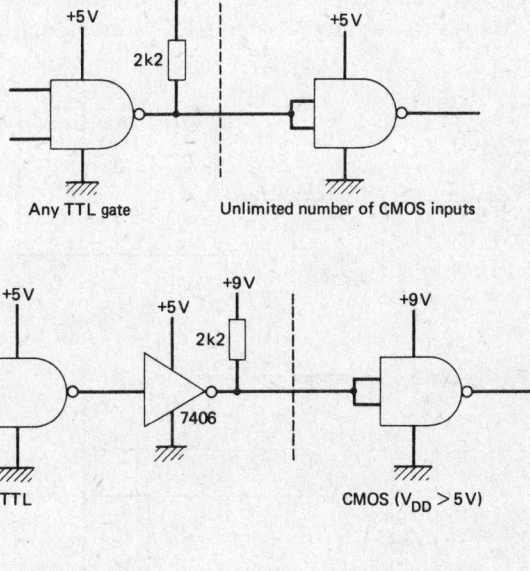

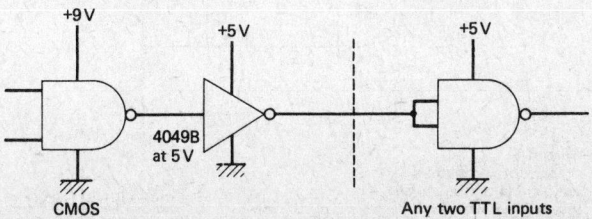

64 Practical Interface Circuits for Micros

3.7 Tri-state Logic   This is not the name for yet another family of logic but a description of logic gates which have outputs that can be connected to a common signal line along with the outputs of other similar gates. With most logic (TTL and CMOS), connecting gate outputs together, apart from not producing the required logic result, would probably damage the ICs. When one output was low while others were high, the low would override the other states and an overload current would be passed. If TTL gates are used which are "open-collector" (i.e. no top transistor in the totem pole output stage), then several such gate outputs can be connected together with a common pull-up resistor to give what is called the wired-or facility. But this will restrict the maximum switching speed of the system. **Tri-state logic** overcomes the problem by having *three* possible output conditions:

a) Low state = Logic Ø
b) High state = Logic 1
c) A high impedance state which effectively open circuits the output.

A tri-state gate has to have an extra input (see *fig. 3.15*) called the control or **enable**. This either allows the gate to operate normally to give a logic state output (Ø or 1) or causes the gate ouput to go into the high impedance state. Several tri-state gates can therefore be used to drive a common line with only one tri-state gate being on at any one time. Gates like these are used extensively in micro systems for bidirectional buses.

Fig. 3.15 Tri-state buffer

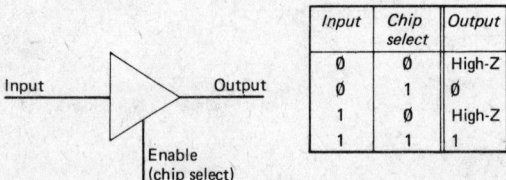

Input	Chip select	Output
Ø	Ø	High-Z
Ø	1	Ø
1	Ø	High-Z
1	1	1

Tri-state gates in a bi-directional arrangement

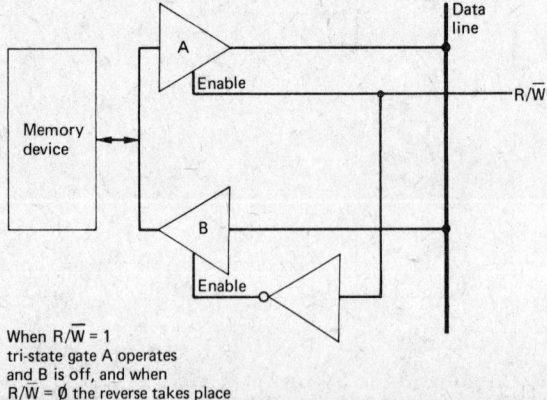

When R/W̄ = 1 tri-state gate A operates and B is off, and when R/W̄ = Ø the reverse takes place

3.8 Bistables

One of the basic requirements within digital systems is for some kind of memory, in other words a circuit that can be set to logic 1 by a pulse or edge and which will hold this level until the circuit gets a reset pulse. The **bistable** is the name given to this type of circuit and it forms the basis for all other types of sequential logic; that is registers, shift registers, counters, dividers, and memories (RAM).

1 As the name implies, the bistable is a circuit which has two possible stable states and the simplest arrangement uses two cross-coupled NAND gates (see fig. 3.16). This gives what is called an **R-S latch**. To SET the circuit, the S input is taken momentarily to logic $\emptyset$. This forces the Q output to be logic 1 and the $\bar{Q}$ output logic $\emptyset$. Because of the cross-coupling, the logic $\emptyset$ from the $\bar{Q}$ output will force the Q output to remain high at logic 1 even when the S input returns high.

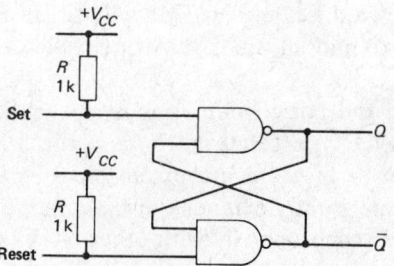

Fig. 3.16 The R-S bistable

Only by applying a logic $\emptyset$ momentarily to the R input will the bistable again change its state. The action is usually described by a truth table, but note that the previous state of the bistable must be included along with the input conditions.

Truth table for the R-S Bistable

Inputs		Previous state of Q	Resulting state of Q	
R	S	Q_n	Q_{n+1}	
$\emptyset$	$\emptyset$	$\emptyset$	X	indeterminate
$\emptyset$	$\emptyset$	1	X	state
$\emptyset$	1	$\emptyset$	$\emptyset$	a $\emptyset$ on the R
$\emptyset$	1	1	$\emptyset$	forces the Q to $\emptyset$
1	$\emptyset$	$\emptyset$	1	a $\emptyset$ on the S
1	$\emptyset$	1	1	forces the Q to 1
1	1	$\emptyset$	$\emptyset$	no change
1	1	1	1	

You will notice from this that the R-S has a basic limitation: for if two $\emptyset$s are applied simultaneously and then removed, the output state will be either a $\emptyset$ or a 1. This restricts the use of the R-S to a few basic applications. *Figure 3.17*

Fig. 3.17 Use of the R-S bistable in switch debouncing

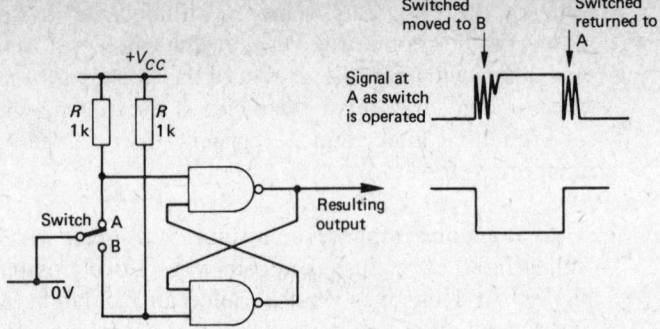

shows one important use of this bistable as a *switch debouncer*. All mechanical switches will "bounce" on make and break, since the contacts open and close several times following a switch operation before finally coming to rest. A logic system will record all these changes as valid inputs. One solution is to wire up an R-S bistable. The bistable will be set or reset at the instant the contacts first make and all other subsequent "bounces" will be ignored.

2 To overcome the basic limitations of the R-S, two other bistables are widely used: the D type (Data latch) and the J-K (usually a master-slave arrangement). Both of these are extensions of a clocked R-S (*fig. 3.18a*). Adding a **clock line** greatly extends the capability of a bistable because it can then be operated synchronously with other circuits.

Fig. 3.18a The Clocked R-S bistable

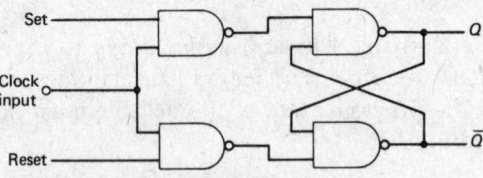

The bistable can be set or reset by logic 1 only while the clock input is high

The **D type bistable** is used for storage of data and has, apart from the clock, one input D for data entry (*fig. 3.18b*). When the clock is high, the state of the D input will be transferred to the Q output, but when the clock is low no change of state can take place. This is called *level clocking*.

Fig. 3.18b The D bistable, or data latch

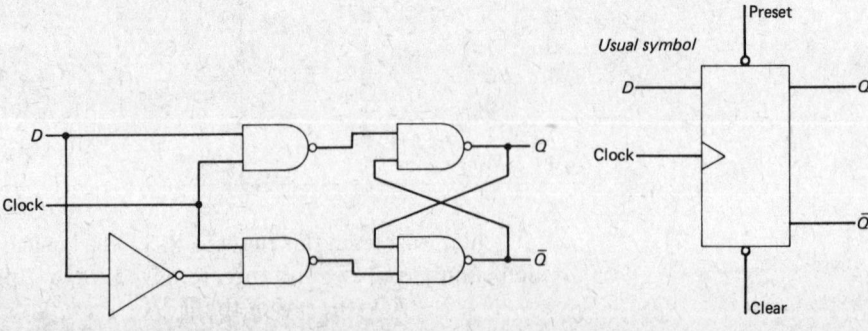

Digital Circuits

Truth table for the D Bistable

Clock	D	Q_n	Q_{n+1}
H	0	0	0
H	0	1	0
H	1	0	1
H	1	1	1

With many IC bistables (TTL and CMOS), the Q output can only change state on *one* edge of the clock waveform. This is called **edge triggering** and has the advantage that data is "locked out" at all times except for the brief instant when the clock changes state.

Several D bistables can be connected together to make a **register** (*fig. 3.19a* and *b*). In its simplest form this gives a *buffer* arrangement (*fig. 3.19a*). The data presented on the register inputs from the bus will be loaded in when a suitable clock pulse is present. This forces the bistables to assume the new state which will be held for the time interval between clock pulses. A system such as this is used in digital multi-meters to ensure that the display is held steady between samples and can also be used as an interface buffer to retain data which is briefly output from a micro.

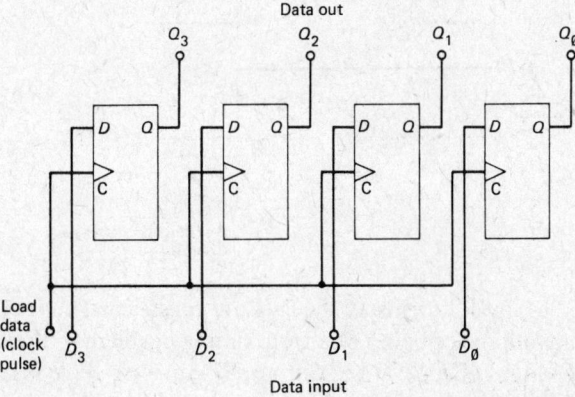

Fig. 3.19a Using D bistables to make a buffer store (only 4 bits are shown but this can be easily extended)

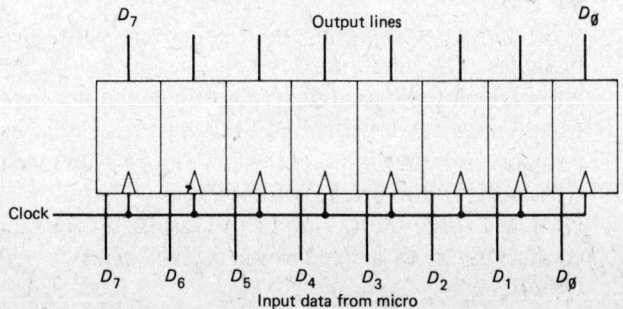

Fig. 3.19b An 8-bit register using D bistables

3 The other very useful bistable is called the **J-K bistable** (*fig. 3.20a*). It is a clocked bistable but, unlike the D, has two inputs. The J-K is therefore similar to the R-S except that it does not have any indeterminate output conditions. This is achieved because of the internal feedback from the Q output which gates the K input and from the $\bar{Q}$ output which gates the J input. When the J and K inputs are both $\emptyset$, there is no change of state, and if they are both logic 1 the output complements.

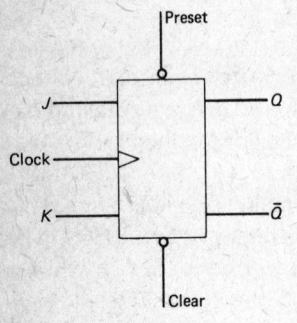

Fig. 3.20a The J-K bistable

Truth table for the J-K Bistable
(with all states it is assumed that a clock signal is applied)

J	K	Q_n	Q_{n+1}	
$\emptyset$	$\emptyset$	$\emptyset$	$\emptyset$	$J = K = \emptyset$
$\emptyset$	$\emptyset$	1	1	no change
$\emptyset$	1	$\emptyset$	$\emptyset$	$J = \emptyset, K = 1$
$\emptyset$	1	1	$\emptyset$	Q forced to $\emptyset$
1	$\emptyset$	$\emptyset$	1	$J = 1, K\emptyset$
1	$\emptyset$	1	1	Q forced to 1
1	1	$\emptyset$	1	$J = K = 1$
1	1	1	$\emptyset$	Q complements

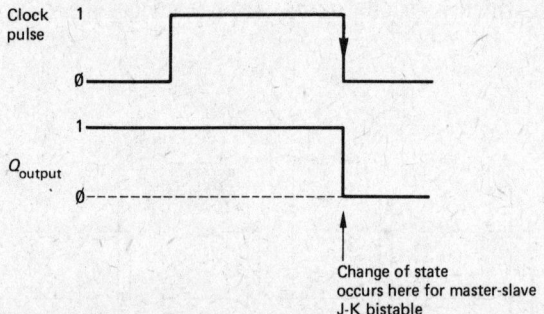

Fig. 3.20b Waveform diagram showing change of state at output with clock pulse (assuming that J and K are each connected to logic 1)

Many IC forms of the J-K are **master-slave** types. These have the distinct advantage of being free from timing problems—the sort that might cause false changes of state at the output. The master-slave is really two bistables in one, but remember when using it that the "slave", to which the Q and $\bar{Q}$ outputs are connected, only changes state on the trailing edge of the clock pulse (see fig. 3.20b).

So far we have concerned ourselves with the R-S and then clocked bistables. The addition of the clock gives the bistable a lot of flexibility and allows them to be used in synchronised circuits. However, these bistables are usually provided with PRESET and CLEAR inputs. These inputs are asynchronous and will therefore operate without a clock signal being present.

A signal to the preset will force the Q output to logic 1 and a signal to the clear will force the Q output to logic $\emptyset$. If these inputs are not to be used, connect them to a fixed voltage level, usually +5 V for TTL and 0 V for CMOS.

3.9 Shift Registers

We have already seen that a register is simply a group of bistables linked together to make some sort of temporary data store. Microprocessors themselves are built using arrays of special registers each performing a separate set of functions described by the instruction set. For example, for an Accumulator, one of the instructions which causes a change in its contents is LSRA (Logical Shift Right Acc A). This has the effect of shifting the contents of the register, that is the state of the bistables, one place to the right—a useful instruction if you just want to separate out and use only the upper 4 bits of an 8-bit word (use LSR four times), but how is this achieved in practice? First of all we have to realise that there are four possible types of **shift register**:

a) Serial-in/serial-out (the simplest, *fig. 3.21a*)
b) Serial-in/parallel-out
c) Parallel-in/serial-out
d) Parallel-in/parallel-out.

Take the circuit of *fig. 3.21b*, a 4-bit parallel-in/serial-out shift register. A new data word can be written into the register when the write line is high and the shift input low. The inputs operate on the preset and clear terminals of the four bistables. Imagine we load in the data 1Ø11. If the write line is then returned low and the shift line pulsed, the state of the bistables (see J-K truth table) moves one place to the right so that, after the shift pulse, the contents read Ø1Ø1. Shift registers are very useful in converting data from parallel to serial and vice versa.

Another application of shift registers is as *sequence generators*, circuits that

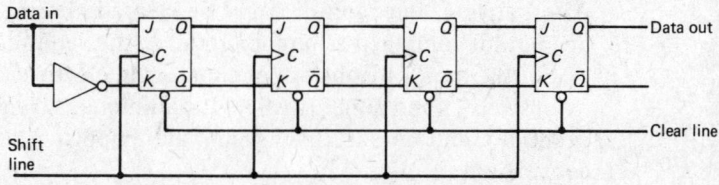

Fig. 3.21a A 4-bit serial-in/serial-out shift register

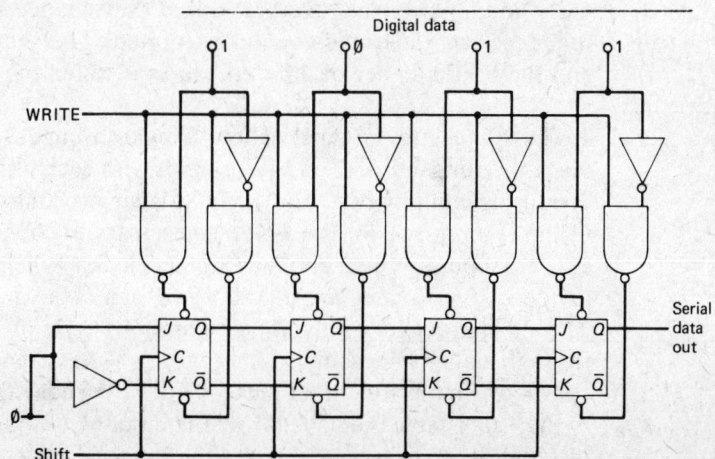

Fig. 3.21b A 4-bit parallel-in/serial-out shift register

Fig. 3.21c A simple sequence generator

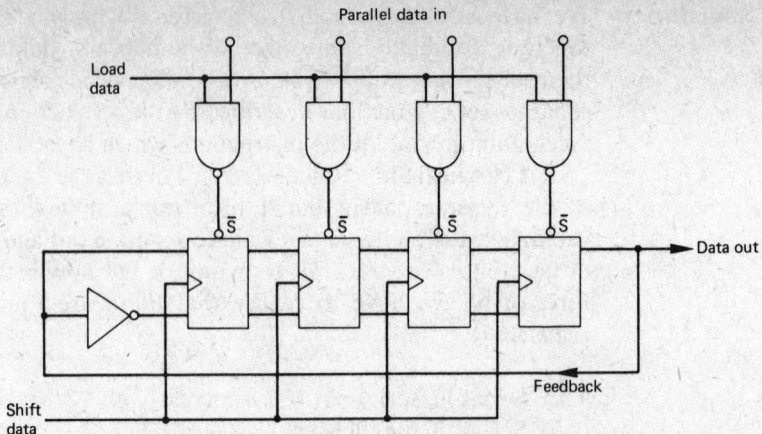

produce a fixed pattern of digital data. A simple example based on the last circuit uses feedback so that the register forms a recirculating store. The required sequence is set up by the parallel inputs and this will appear at the ouput and be repeated every four shift pulses (*fig. 3.21c*).

3.10 Counters

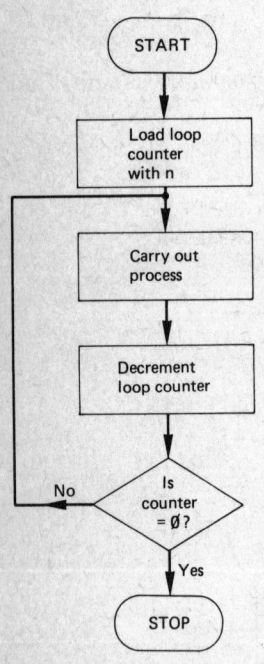

Fig. 3.22 Typical structure of a flow chart containing a loop

Anyone who has done even a limited amount of programming will have come across the use of **counters**. In a program with a loop, one of the registers inside the microprocessor will be set up as a *loop counter*. Initially it will be loaded with the number of times that the process has to be repeated, i.e. the number of passes, and then decremented (i.e. counted down) for each pass until zero is indicated. A typical flowchart for this is shown in *fig. 3.22*. Alternatively, the counter could be cleared and then incremented (counted up) until it contained a number equal to the required value of passes. This is one of the most obvious use of counters in electronics, but they are also used extensively in instruments (digital multimeters, digital frequency meters, for example) and in interfacing for the important task of analog-to-digital conversion.

By linking bistables together so that they change state in a predetermined sequence, an electronic counter is formed. The sequence is called the *code* and the total number of different states is called the *modulo* of the counter.

1 The simplest of circuits, called **"ripple-through"** counters (*fig. 3.23*), are made up using either D or J-K bistables with each bistable connected so that it divides its input by two. For the D, this means connecting the $\bar{Q}$ output back to the D input and for the J-K it is necessary to connect both J and K to logic 1. The input is then applied to the clock. When several of these basic divide-by-two circuits are linked up we can make a $\div 4$, $\div 8$, $\div 16$ and so on. The only problem with this **asynchronous** type of counter is that a ripple through delay builds up so that, for a $\div 16$ arrangement, as the counter overflows on the 16th input pulse the negative edge that appears at the output of the 4th bistable is delayed from the input by four propagation delay periods (see *fig. 3.23*).

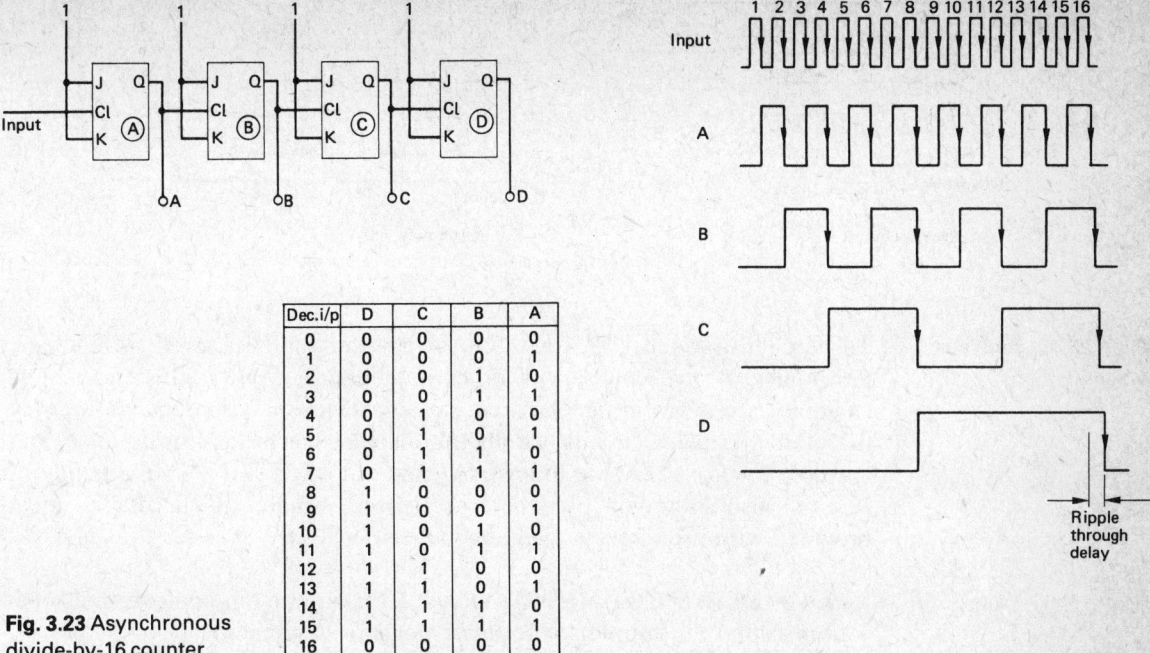

Dec.i/p	D	C	B	A
0	0	0	0	0
1	0	0	0	1
2	0	0	1	0
3	0	0	1	1
4	0	1	0	0
5	0	1	0	1
6	0	1	1	0
7	0	1	1	1
8	1	0	0	0
9	1	0	0	1
10	1	0	1	0
11	1	0	1	1
12	1	1	0	0
13	1	1	0	1
14	1	1	1	0
15	1	1	1	1
16	0	0	0	0

Fig. 3.23 Asynchronous divide-by-16 counter

2 Synchronous counters (*fig. 3.24*) overcome the problem of cumulative delay by ensuring that, when bistables have to change state, they all change state at the same instant. In a synchronous ÷16 (8421) binary counter, the Q output of each bistable is connected to the J and K inputs of the next and so on, and all clock inputs are connected together. Extra AND gates are required to ensure that the counter changes state in the required sequence. In this way bistable C can only change state when both bistables A and B are at logic 1; this occurs after the 4th input pulse. Similarly, bistable D can only change state after the 8th input pulse because this is the first time that the Q outputs of bistables A, B and C are at logic 1. Synchronous counters allow dividers of large numbers to be created without generating long delays and are less prone to produce "glitches" when the circuit is decoded.

Fig. 3.24 Synchronous divide-by-16 counter

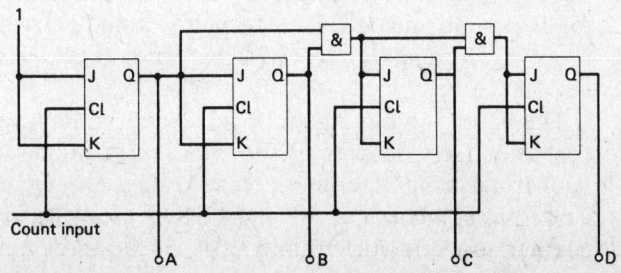

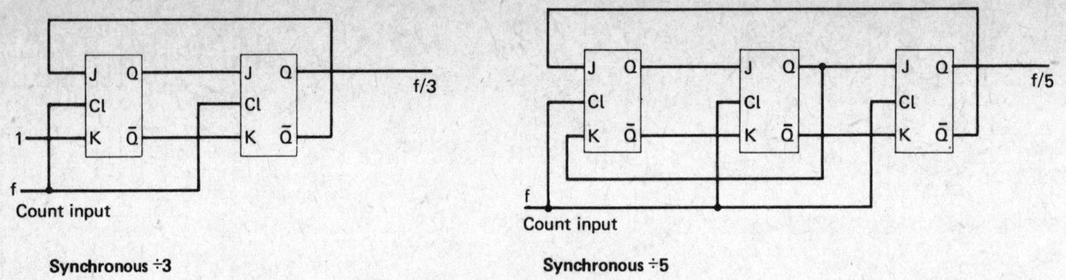

Fig. 3.25 Synchronous counters

3 The count sequence of a pure binary counter can be altered by feedback techniques to give counters of other numbers, 3, 5, 7, 9, 10, etc. A few examples are given in *fig. 3.25*. As stated previously, counters can be very useful in interfacing circuits, for **digital-to-analog convertors** for instance, and a simple low-cost example follows. A glance at the TTL or CMOS family of ICs will also show you that there are several counter ICs available, most provided with useful extra facilities such as

a) UP/DOWN COUNTING A logic level on one pin controls the direction so that the counter either increments or decrements.
b) PRESETTABILITY A data word can be loaded into the counter so that it starts counting from that value.
c) CLEAR The correct pulse clears all the internal bistables so that they all hold $\emptyset$.
d) BCD/BINARY COUNT A logic level applied to this pin causes the counter to either advance in a pure binary sequence or act as a decade ($\div 10$) counter.

In *fig. 3.26*, four D bistables (2 CMOS 4013 ICs) are wired up as a $\div 16$ counter and the Q outputs are connected to an R-2R ladder network. When clock pulses are applied the counter advances and the R-2R resistor network converts the various states into a ramp-type output. The circuit is therefore a type of digital-to-analog convertor. If, for example, only one or two spare output ports from the micro system are available, it is possible to make a type of serial digital-to-analog convertor from this circuit. Initially the counter would be cleared by pulsing the reset line and then a fixed number of pulses would be directed to the counter from the main output port. The number of pulses from the micro under software-controlled output would be equivalent to the desired analog level. The pulses would be counted and the state of the counter decoded by the R-2R ladder network to give the analog output voltage.

The basic circuit of *fig. 3.27* can also form the heart of an analog-to-digital convertor (see Chapter 5). For this, in addition to the counter and R-2R network, a comparator and a few CMOS gates (*fig. 3.27*) will be needed. The clock pulses generated by the CMOS gated oscillator are applied to the counter just after the trailing edge of the reset pulse. The ramp output is

Fig. 3.26 Divide-by-16 counter circuit using four D bistables

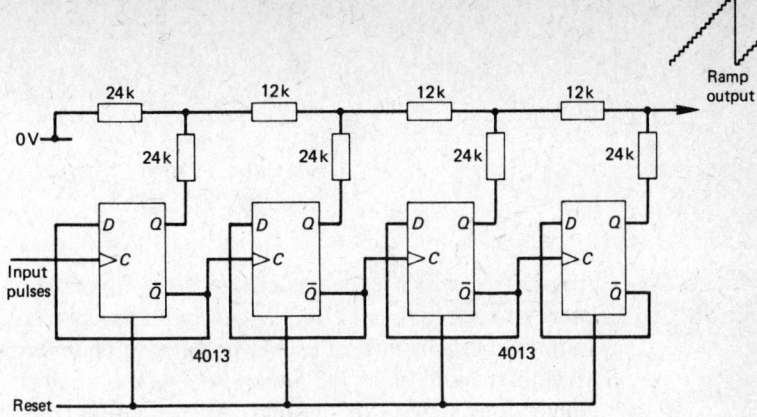

Fig. 3.27 A basic analog-to-digital convertor

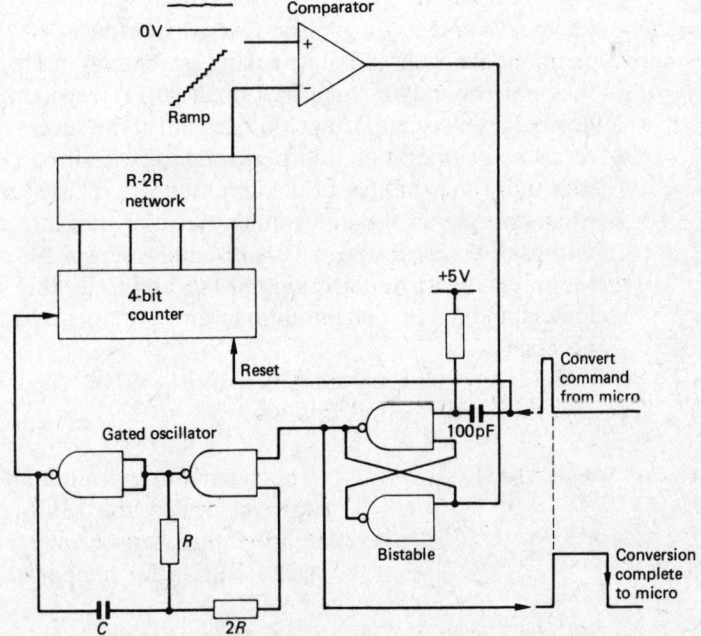

compared with the analog input and, when the ramp just exceeds the analog value, the comparator output switches low and resets the control bistable to stop the clock. The total count achieved by the counter (4 bits in this case) will then be equivalent to the analog input.

4 Sensors and Output Devices

4.1 Transducers— an Introduction

Any device which converts energy from one form to another, such as light energy to mechanical energy, can be called a transducer. For our types of system, the transducers used will have to change input energy (heat, light, movement) into electrical energy to give the small signals which, after A-to-D conversion, actuate the system. At the output, this process is reversed, with electrically driven power transducers providing the required output energy (*fig. 4.1*).

Typical input transducers (SENSORS) are devices like thermocouples, photocells and strain gauges. Output transducers (OUTPUT DEVICES) are components such as loudspeakers, solenoids, motors and valves. Although they may be only a small part of the overall system, the sensors and output devices are very important. In particular, the success of any control system in terms of its operation and performance will often depend upon the quality, sensitivity and stability of the input sensors. These sensors have to pick up the small changes in the input quantities and translate these often tiny changes into useful electrical signals. Choosing a sensor for a system that is right for the job is therefore fairly significant, and to do this with some confidence an understanding of the limitations and parameters of the various devices is necessary.

The important parameters which will be used in the specifications of sensors are defined as follows.

- RANGE — The maximum and minimum values of input quantity over which the sensor is usable. Example: a temperature sensor may be specified as giving a useful output for temperatures in the range 20°C to +400°C.
- ACCURACY — The degree to which the output of a sensor is in agreement with the true value. In other words, the "error" between the indicated and the correct value. Note that this parameter can be expressed in several ways such as ±1% of reading or ±1% of full scale reading (output). The first gives the tightest tolerance.
- SENSITIVITY — This refers to the change in output for a unit change in input quantity. Example: a thermocouple may have a sensitivity of 40 μV per °C. In most cases, the highest sensitivity possible will be required.

Fig. 4.1 Input and output transducers in a micro-based system

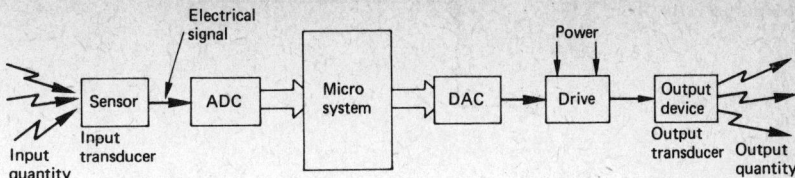

- RESOLUTION The smallest change in the input quantity which will give a measurable output. Some devices such as wire-wound resistive tracks give outputs that increase in a step-like fashion. With these the resolution is determined by the number of turns.

- LINEARITY Over the useful range of the sensor, the change in output for corresponding changes in input quantity should be uniform if perfect linearity is required. In practice some nonlinearity in a sensor will always exist. In many cases nonlinearity can be compensated for by software techniques.

- REPEATABILITY Suppose at a light level of 50 Lux ($0.25\,\text{mW/cm}^2$) a photoconductive cell gives a resistance reading of $2200\,\Omega$. If the light level is changed several times to higher or lower values, but always returned exactly to 50 Lux, a series of resistance readings for the input at 50 Lux could be taken. The variations in the readings or "scatter" (i.e. $2204\,\Omega$, $2185\,\Omega$, $2218\,\Omega$, etc.) would represent the "repeatability" of the sensor. Repeatability is therefore a measure of how near the output of a sensor returns to the same output when measuring the same input value.

- RESPONSE TIME This is the time taken for the sensor output to change and settle to a new value following a step-like input change. There is a variation in the way in which this parameter is specified and measured. If fast response from a sensor is required, especially for very large changes in input quantity, then the specification for the sensor must be studied carefully (fig. 4.2).

- STABILITY The output level from a sensor will drift even when the input is held constant. The drift will be caused by ageing and by environmental changes. Temperature changes in particular will cause most sensors to drift, and the temperature characteristic or temperature coefficient may well be quoted separately.

The list given above is not necessarily complete and for any particular application you should investigate the fixing or mounting method for the

Fig. 4.2 Response time of a sensor

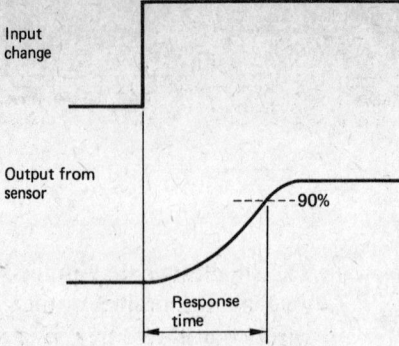

sensor and check for adverse environmental conditions such as: high levels of humidity; mechanical shocks and vibrations; susceptibility to electrical noise, and so on. It is as well to choose a device that is robust and well shielded if satisfactory long-term operation is required.

The following sections deal with commonly used sensors and output devices together with some application information. The types discussed represent only a small part of the large numbers available but will give a guide to the general principles involved.

4.2 Temperature Sensors

Next to the design job itself, the task of preventing excessive changes in circuit performance with temperature has always been a concern of electronic engineers. This is because temperature variations can cause large changes in component values and parameters and consequently to the operating performance of some unit. On the other hand, by exploiting those very changes in the electrical properties of components, it has been possible to create a wide variety of useful temperature sensors. Some of the popular types are listed in the table:

Sensor	Features	Typical useful temperature range
THERMISTOR	Resistance falls with temperature.	−80° to +300°C
THERMOCOUPLE	Voltage output rises with temperature.	0° to +1000°C
PLATINUM RESISTANCE	Resistance rises with temperature.	−50° to +500°C
SEMICONDUCTOR TYPES	Voltage for pn junction falls by 2 mV/°C. Current output rises with temp.	−55° to +150°C

1   A **thermistor**, made basically of sintered oxides of nickel and manganese, is a device that undergoes a very large change of resistance with temperature. The name is derived from *therm*ally sensitive res*istor*. Most of the types have a negative temperature coefficient (NTC) but some positive temperature coefficient varieties (PTC) are also available. The material can be formed into rods or small beads, but for sensing purposes the small bead shape is used in order to get the fastest possible response. The bead is then sealed inside a glass envelope or into a stainless steel probe (*fig. 4.3a*).

The resistance follows the law:

$$R_2 = R_1 e^{B(1/T_1 - 1/T_2)}$$

where B = characteristic temperature constant (K)
 T = bead temperature (K)
 R_1 = resistance of thermistor at temperature T_1
 R_2 = resistance of thermistor at temperature T_2
 e = 2.7183.

For a device such as the GM102, which has a quoted resistance at 25°C of 1 kΩ, $B = 3000$. Therefore, at +100°C (the maximum temperature for this thermistor is 125°C),

$$R_2 = 1000 e^{-2.018} = 133\ \Omega$$

By working out values for other temperatures a graph can be plotted as in *fig. 4.3b*. Note that the formula gives a straight line when a logarithmic scale is used for resistance values.

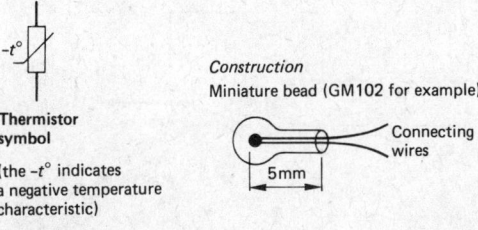

Thermistor symbol

(the $-t°$ indicates a negative temperature characteristic)

Fig. 4.3a Thermistor symbol and construction

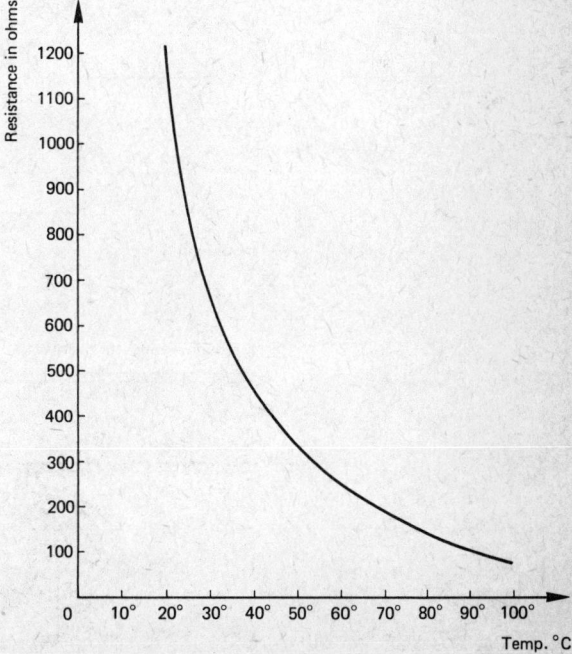

Fig. 4.3b Graph of thermistor resistance against temperature

Fig. 4.4 Thermistor data (miniature beads)

	GM 102	GM 472	GM 473
Resistance 20°C	1 kΩ	4k7 Ω	47 kΩ
R_{min} (hot)	59 Ω	271 Ω	338 Ω
Tolerance	±20%	±20%	±20%
Temperature range	−80°C to +125°C	−80° to +125°C	−60 to +200°C
Maximum dissipation	70 mW	70 mW	120 mW
Time constant (thermal)	5 sec	5 sec	5 sec
Temp. constant B	3000	3390	3930
*Dissipation constant	0.7 mW/°C	0.7 mW/°C	0.7 mW/°C

*Dissipation constant refers to the self-heating effect and it represents the amount of power required to raise the temperature of the thermistor 1°C above ambient.
  For example, suppose $I_t = 0.1$ mA and $R_t = 10$ kΩ.
  Power dissipated by the thermistor is

$$P_t = I_t^2 R_t = (0.1 \times 10^{-3})^2 \times 10^4 = 0.1 \text{ mW}$$

In this case the dissipation would only cause a very slight difference between the temperature of the thermistor and that of the surroundings.

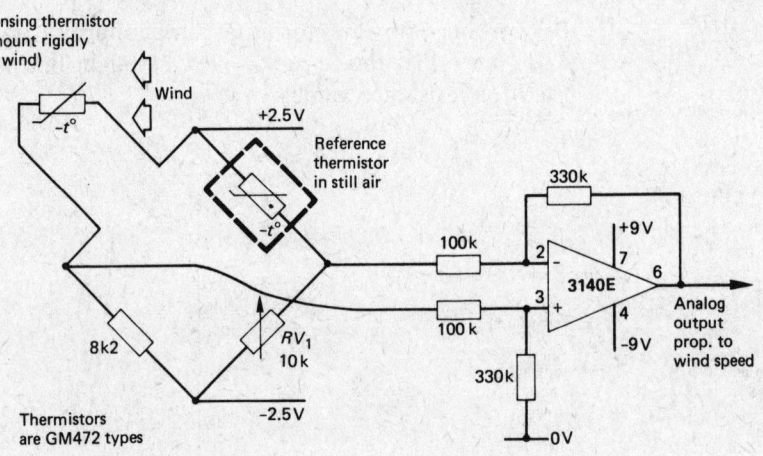

Fig. 4.5 Thermistor in a bridge arrangement

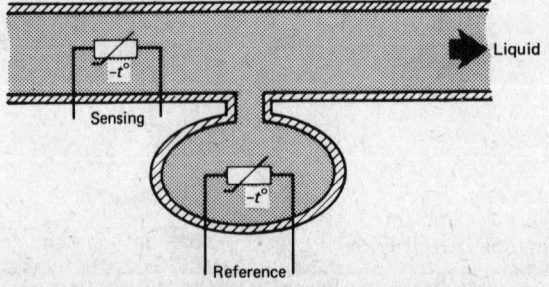

Alternative sitings for thermistors

We have already covered one application of the thermistor as a sensor in the oven temperature control example given in Chapter 1. In that arrangement the thermistor was supplied from a constant-current circuit so that, as the temperature varied, causing a change in thermistor resistance, so the output voltage varied. One important point to remember in any arrangement is that the current flowing in the thermistor must not be too large otherwise a self-heating effect will take place, causing some inaccuracy in the output. In no situation should the maximum dissipation (power = current^2 times resistance) be exceeded. Data for three types is given in *fig. 4.4*.

The other circuit to use with a thermistor is the Wheatstone bridge. *Fig. 4.5* shows a typical bridge circuit which in this case is used for sensing wind speed or liquid flow. A reference thermistor in still air (or in a chamber out of the liquid flow) is fixed in one arm of the bridge, while the sensing thermistor in the other arm is mounted to pick up wind speed (or liquid flow). Under zero wind conditions, the bridge is balanced by RV_1 so that the output from the amplifier is also zero. When the sensing thermistor is then exposed to the environment, any wind will cool the sensing thermistor more than the reference thermistor, the bridge will be unbalanced, and an output voltage will be provided from the amplifier. This voltage will be proportional to wind speed (or liquid flow).

2   A **thermocouple** is a sensor which consists of two junctions of dissimilar metals. The operation, based on a discovery in 1921 by Seebeck, is that an e.m.f. will be generated proportional to the temperature difference between these two junctions (*fig. 4.6*). One junction is used for measuring temperature while the other is used as a reference. The reference is normally maintained at a constant temperature, say 0°C, or in a temperature-controlled enclosure at a value slightly higher than ambient.

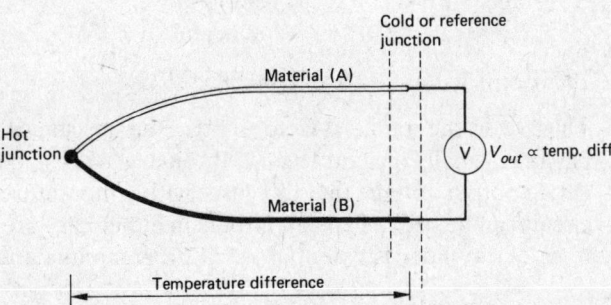

Fig. 4.6 Principle of operation of the thermocouple

Alternatively, variations in temperature at the reference can be compensated for by a thermistor. The commonly used thermocouple materials are

Materials	Working temperature range
Nickel Chromium with Nickel Aluminium	−50°C to +400°C
Copper with Constantan	−250°C to +400°C
Iron with Constantan	−200°C to +1200°C
Platinum with Platinum and 13% Rhodium	−50°C to +1750°C

The output of thermocouples is not high—a typical value is 40 μV per °C—so an amplifier is essential to boost the signal before applying it to an ADC. However, thermocouples have the advantage of being linear devices with good accuracy and high stability. A typical circuit using a NICR/NIAL type is shown in *fig. 4.7*.

3 **Platinum**, an inert stable material, exhibits a known and repeatable change of resistance with temperature. It can therefore be used as an accurate temperature sensor over a wide range without the need for junction compensation. The sensor is made either using a coil of platinum wire on an insulating former or as a film of platinum deposited onto an alumina substrate (*fig. 4.8*).

Typical data is:

Resistance at 0°C	Temperature coefficient	Temperature range
100 Ω ± 0.1 Ω	+0.385 Ω/°C	−50° to +500°C

Thus over the range 0°C to 100°C, the resistance changes by 38.5 Ω. By passing a small constant current through the sensing film it is then possible to get an output voltage that increases with temperature. Alternatively, a bridge circuit can be used. Typical outputs in either case are about 1 mV/°C so some form of amplifier is essential. Measuring circuits are shown in *fig. 4.9*.

4 The leakage current of **semiconductor** materials such as silicon increases with temperature. This means that the voltage required to forward bias a pn junction (a diode or the base/emitter junction of a transistor) also falls with temperature. Typically this fall is about 2 mV/°C (*fig. 4.10*). This characteristic enables a pn junction to be used as a temperature sensor by supplying it with a constant current and detecting the changes in forward voltage.

Fig. 4.7 Basic amplifying circuit

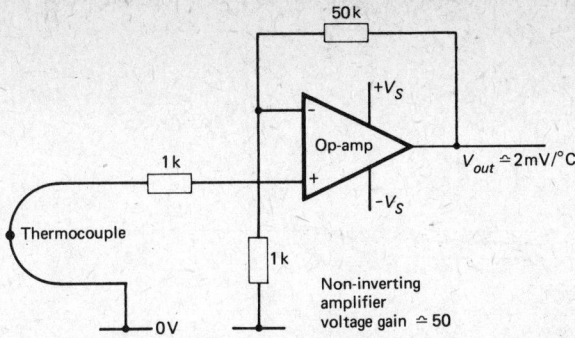

Fig. 4.8 Platinum film resistance temperature detector

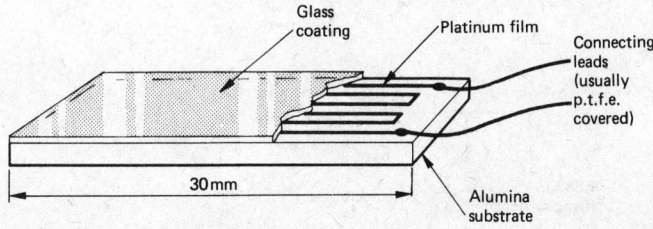

Fig. 4.9 Measuring circuits for use with PRTD

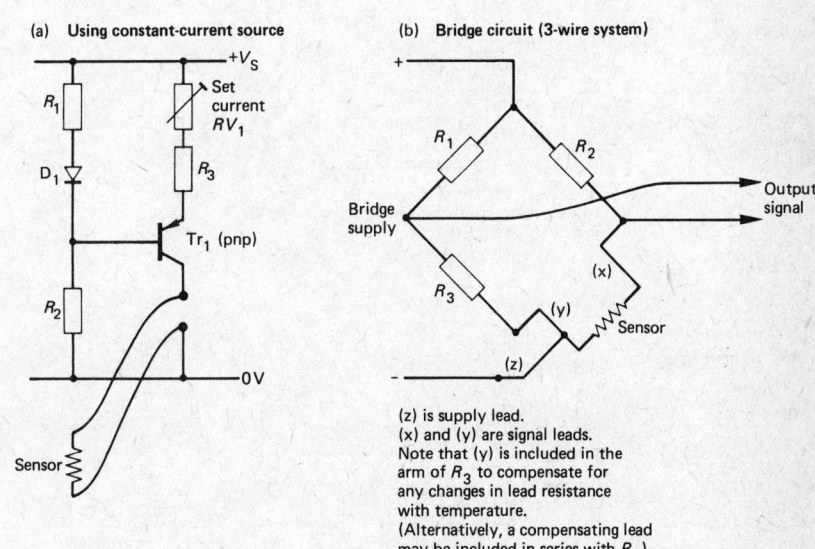

(z) is supply lead.
(x) and (y) are signal leads.
Note that (y) is included in the arm of R_3 to compensate for any changes in lead resistance with temperature.
(Alternatively, a compensating lead may be included in series with R_3).

Commercial semiconductor temperature sensors such as the RS590 utilise the temperature-dependent characteristics and are devices which produce an output current proportional to absolute temperature. Over the supply voltage range of +14V to +30V d.c. the RS590 acts as a high resistance current generator with the value of current being 1 µA per degree Kelvin. The temperature range is from −55°C to 130°C with an accuracy of ±2.0°C.

Fig. 4.10 Variation of pn junction characteristics with temperature

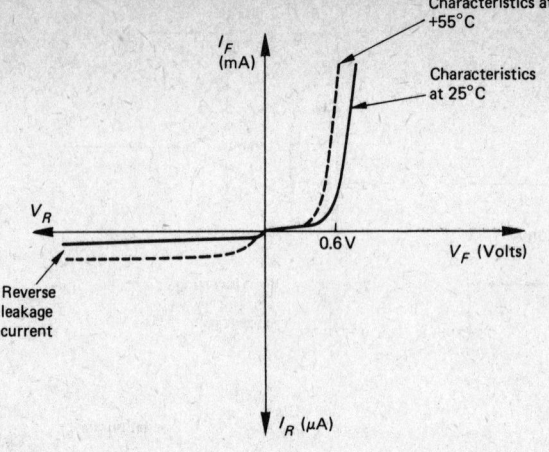

Fig. 4.11a Simple temperature-measuring circuit using the RS590

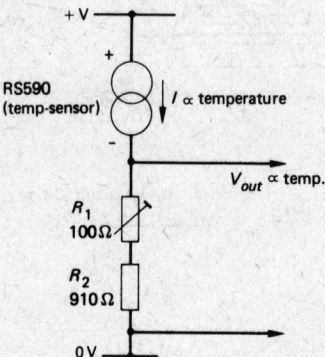

To set output, use known temperature and set R_1 so that V_{out} = 1mV/K.

Fig. 4.11b Using the RS590 with an amplifier to give two-point temperature trim

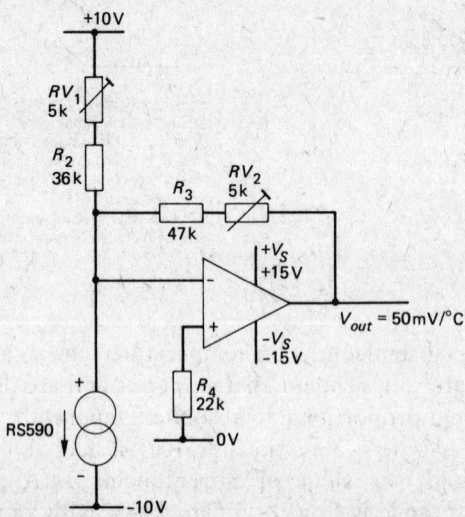

Fig. 4.11c Connections to an amplifier circuit when the temperature sensor is remote

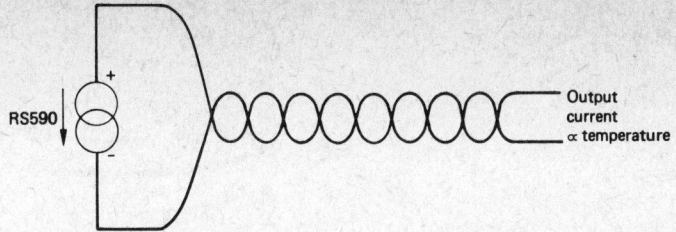

Very simple circuits to sense temperature can be made by placing a resistor in series with the sensor (*fig. 4.11a*). This also shows a method for trimming out any small calibration error. A more complicated circuit with a two-point temperature trim is also shown (*fig. 4.11b*). The circuit is initially adjusted by varying RV_1 to give zero volts output when the sensor is at 0°C. The sensor temperature is then set to +100°C (or other suitable known value) and RV_2 is adjusted to give a full-scale output of say +5 V. One final point is that this type of sensor can be used for remote monitoring purposes. This is because it is a current output device and will therefore be relatively insensitive to voltage drops over long leads. The connection method should be via a twisted pair to avoid pick up of noise (*fig. 4.11c*).

4.3 Light Sensors and Devices

The general name of **opto-electronics** is given to the wide variety of devices used to detect light (photo-sensitive) and to generate and radiate light (photo-emissive). This term for our purposes covers:

Photoconductive cells (light-dependent resistors)
Photodiodes and phototransistors
Photovoltaic cells (silicon solar cells)
Slotted opto-switches
Reflective opto-switches
Light-emitting diodes.

Before discussing these various sensors and devices it will be useful to take a brief look at the units used in illumination. The unit of radiant flux (Φ_e) is the Watt. But for the spectral portion (i.e. the visible part of the spectrum) the unit for flux or luminous power is the Lumen (lm). Luminous power is the total visible light energy emitted by a source in unit time. The unit of illumination (i.e. how much light falls on a given area) is the Lux, or Lumen per Square Metre. Some idea of the levels involved are indicated by the following table:

Light source	*Illumination* in Lux (lumen/sq. metre)
Bright sunlight	25 000
Fluorescent lights	500
60 W lamp at 1 m	50
Moonlight	0.1

Fig. 4.12 Photoconductive cell (light-dependent resistor)

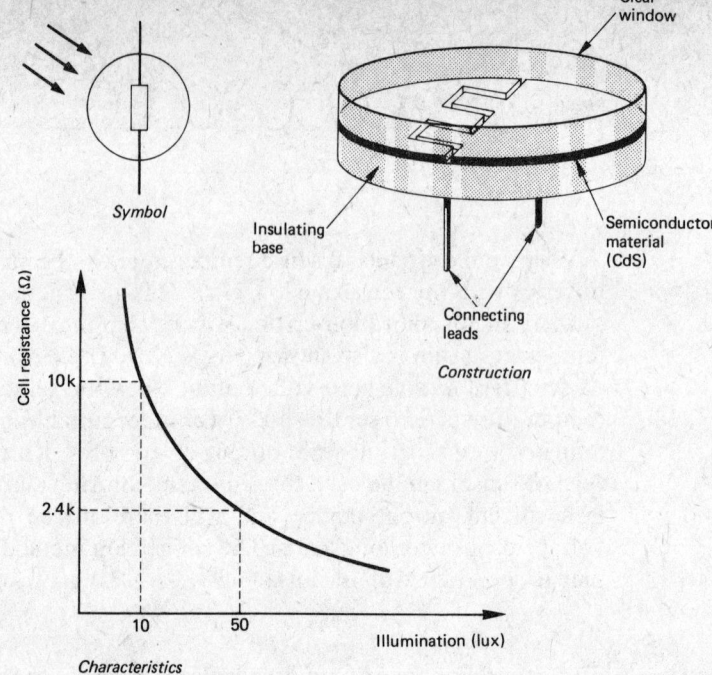

1 **Photoconductive Cells** (*fig. 4.12*)   The resistance of these devices falls as the amount of light incident on a cell increases. For this reason they are often called light-dependent resistors (LDR). The most commonly used types (such as the ORP 12) are made from a film of the semiconductor material cadmium sulphide. The spectral response of a CdS cell closely matches that of the human eye, which is one reason for the wide use of these cells in lightmeters and light-level detectors. The film of CdS is deposited on to an insulating substrate, electrodes are added to the two ends of the cell pattern, and the unit is then encapsulated in epoxy resin leaving a clear end window.

In the dark, a cell like the ORP 12 has a resistance of nearly 10^7 ohms but this falls rapidly with illumination, following a log law as shown. The response time is not fast, with typical resistance rise times of 75 msec and fall times of 350 msec so do not expect an LDR to be able to pick up signals from rapidly moving objects. The device is particularly suited to detecting slowly varying light levels, as smoke detectors and in lighting controllers. The change in resistance with light from the cell can be converted into a voltage by supplying the cell with a constant current or by using a bridge arrangement in the same way as for a thermistor.

2 **Photodiodes and Phototransistors** (*fig. 4.13*)   These are much faster-operating devices than the LDR with rise times of a few microseconds or even nanoseconds in some cases. The photodiode, usually silicon, has a small glass window that allows light to fall on the pn junction. This light creates

Fig. 4.13 Photodiode and phototransistor

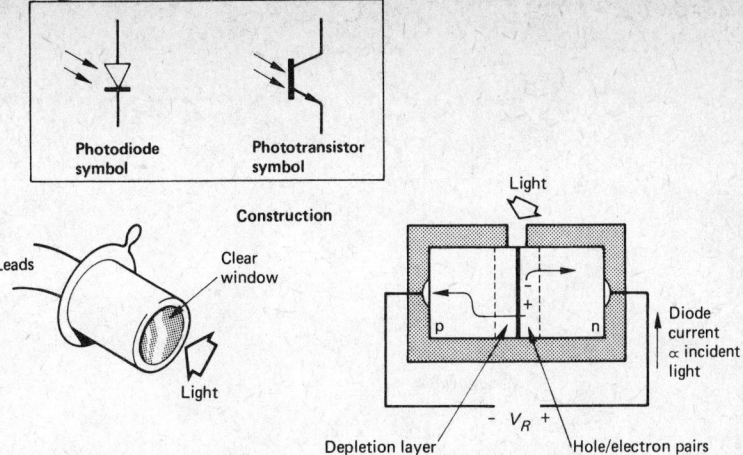

Fig. 4.14 Light-level detector

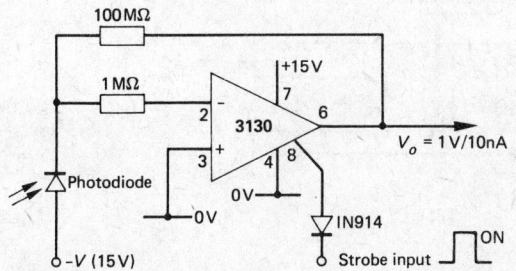

hole-electron pairs at the junction and therefore induces current in the diode. Normally, the diode is operated with reverse bias so that its dark current is very low (typically a few tens of nanoamps with a reverse bias of 20 V). Then, as the incident light is increased, the output current will rise in proportion and this relationship is very linear. Typical sensitivity is about $1\,\mu A/mW/cm^2$. A circuit arrangement for detecting light levels and allowing the output to be strobed via a micro is shown in *fig. 4.14*. The op-amp is a MOSFET type 3130 which has a very high input resistance ($1.5 \times 10^{12}\,\Omega$). With the resistor values shown, the output is 1 V per 10 nA of diode current.

The phototransistor uses the same principle as the photodiode, but because it is an amplifier the device is even more sensitive. Light falling on the base of the transistor via the glass window sets up a base current and this is amplified to appear as a large collector current.

3 Photovoltaic cells These silicon devices, sometimes called *solar cells*, generate an e.m.f. between the output terminals proportional to the incident light. Again it is basically a pn junction, but one of the regions is very thin so that light can pass without too much energy loss. This light causes hole-electron pairs to be generated at the junction. Because one region is thin this rapidly saturates and a voltage is set up. The cell can, of course, be used in the photoconductive mode with current output instead of voltage.

Fig. 4.15 Opto-switches

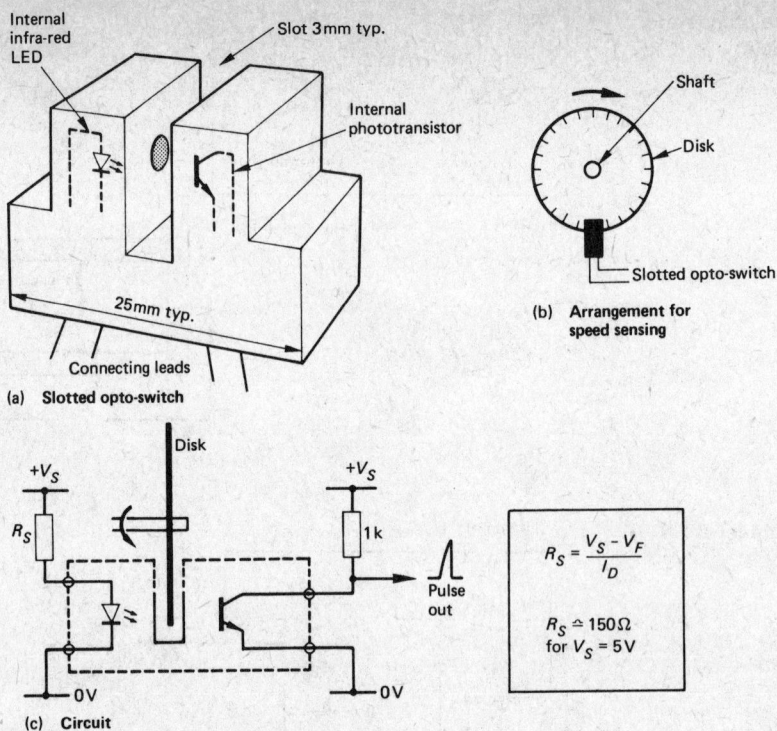

(a) Slotted opto-switch

(b) Arrangement for speed sensing

(c) Circuit

$$R_S = \frac{V_S - V_F}{I_D}$$

$R_S \simeq 150\,\Omega$ for $V_S = 5\,V$

4 Opto-switches (*fig. 4.15*) These sensors, which provide an ON/OFF type of indication, are particularly useful for limit detection, batch counting, position sensing, and level indication. The units are constructed using an infra-red emitting diode and a silicon phototransistor in one moulded package. The optical link in the first type (*fig. 4.15*) is across a slot, and the beam can be interrupted to give a positive output pulse from the phototransistor. Suppose the speed of rotation of a shaft is required. A disk can be fitted to the shaft as shown and the slotted opto-switch can be positioned so that the disc revolves through the slot. The disk will be marked with dark bands and, as it revolves through the slot, the infra-red beam will be blocked and the phototransistor will switch off to give positive output pulses. The number of pulses per second is proportional to the speed of rotation. A similar arrangement can be used for position sensing; in fact there are almost an endless variety of applications.

The reflective opto-switch responds only to the presence of a light-reflecting surface between the infra-red LED and the photo-Darlington. The distance for optimum response is typically 4 to 5 mm. In the circuit example (*fig. 4.16*) the external transistor Tr_1 will be on until the reflective surface completes the optical link. When this occurs, the photo-Darlington conducts, diverting base current away from Tr_1, causing the output to switch high. A device like this is ideal for detecting or counting reflective targets. Note that it is important that a correct-value resistor is wired in series with the LED. This applies to all LED devices and the resistor should be chosen so that the diode current is set to about 20 mA (absolute maximum is usually 40 mA). If the supply is +5 V then R_S is about 150 Ω. Watch also that the reverse voltage rating (2 V) of the LED is not exceeded. This can happen if the supply is accidentally reversed.

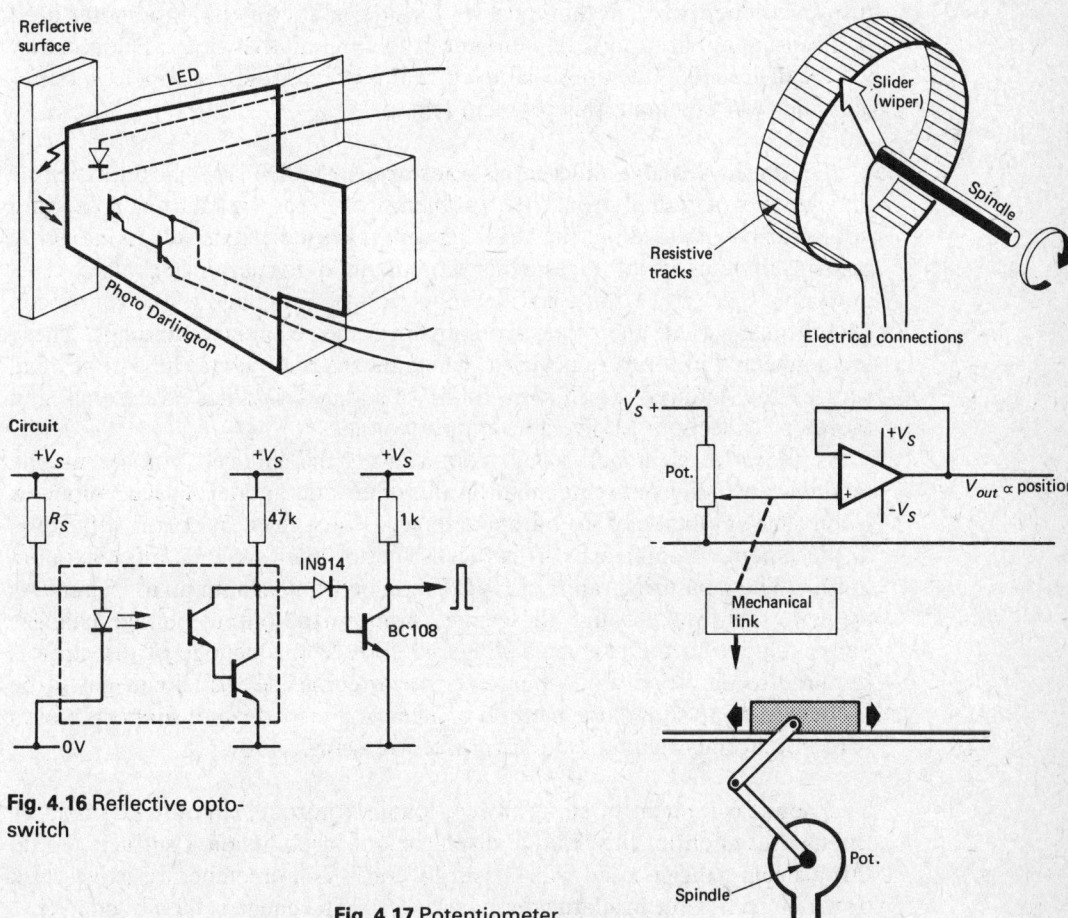

Fig. 4.16 Reflective opto-switch

Fig. 4.17 Potentiometer

4.4 Position and Force Sensors

A whole range of devices is available for sensing angular position, linear displacement, force and pressure.

1 The resistance **potentiometer** is one of the simplest types of position sensor. In its basic form (*fig. 4.17*) it consists of a linear resistive track on which a sliding contact (called the wiper) rotates.

The track of the potentiometer is connected across a d.c. supply and the spindle is mechanically linked to the movement being measured. The output voltage from the wiper will then be directly proportional to the movement. If continuous action is expected then the resistive track and the wiper will be subjected to wear. All potentiometers have a restricted working life which is usually quoted as the total number of complete rotations that can reasonably be expected before failure. A typical figure for a wire-wound potentiometer is about 50 000 rotations. Wire-wound tracks are often preferred to carbon but the voltage output from a wire-wound changes in small steps as the slider moves from turn to turn. The resolution depends upon the total number of turns. Another problem that must be considered for any potentiometer used in a sensing application is the amount of loading on the wiper. The linearity will be degraded if any appreciable current is taken from the wiper. A buffer amplifier will eliminate this problem (*fig. 4.17*).

2   The **linear variable differential transformer** (LVDT) is a position sensor that does not suffer from the problems of wear, resolution loss, and nonlinearity with loading. For these reasons it is used extensively in industrial control. It consists of three coils wound on a former inside which is a moveable core (*fig. 4.18a*). An a.c. input voltage is applied to the centre coil (the primary) while the other two coils form the secondary windings. These are connected in series opposition, which means that, when the core is dead centre, the induced e.m.f.s in both secondary windings are equal and therefore cancel out to give zero output voltage.

As the core is moved away from centre, the induced voltage in one secondary winding is greater than in the other and an a.c. output voltage is given. The amplitude of the output voltage is a very linear function of the core displacement. Nonlinearity is typically better than ±0.5% of maximum output. The measuring range of LVDTs can be from 0.1 mm up to 75 mm. As the core is moved through the centre position, the output voltage changes phase relative to the primary voltage by 180°. If this change of direction is required to be picked up, a phase sensitive rectifier (PSR) circuit has to be used. *Fig. 4.18b* shows one method of achieving a d.c. output proportional to core displacement either side of zero centre using a PSR.

3   The measurement of strain, force, load and torque can only be made by the measurement of the relative displacement of points on a surface. To do this a **strain gauge** is used. This basically consists of an etched resistive track or wire on a flexible insulating base (*fig. 4.19*). The gauge is then bonded (i.e. cemented) to the mechanical member in which the strain is to be measured. The principle of operation depends on the fact that the resistance of an element changes when its dimensions are altered. Therefore, if the member to which the gauge is bonded is stressed, the strain caused can be measured by recording the change in resistance of the gauge. The formula for the resistance of a thin wire or film is

$$R = \rho \frac{l}{a}$$

where ρ is resistivity
l is length
a is cross-sectional area.

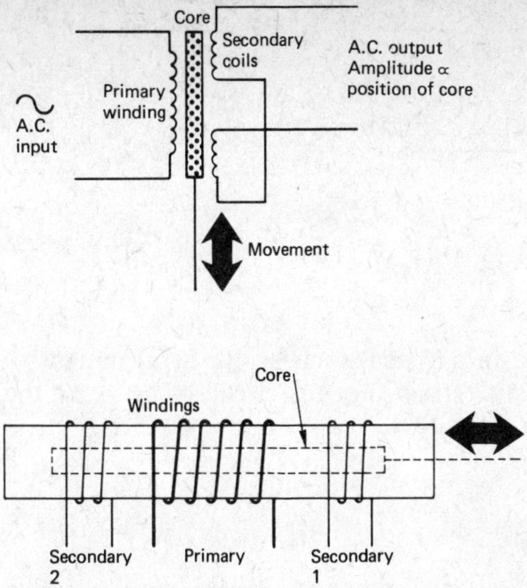

Sensors and Output Devices 89

Fig. 4.18a Linear variable differential transformer

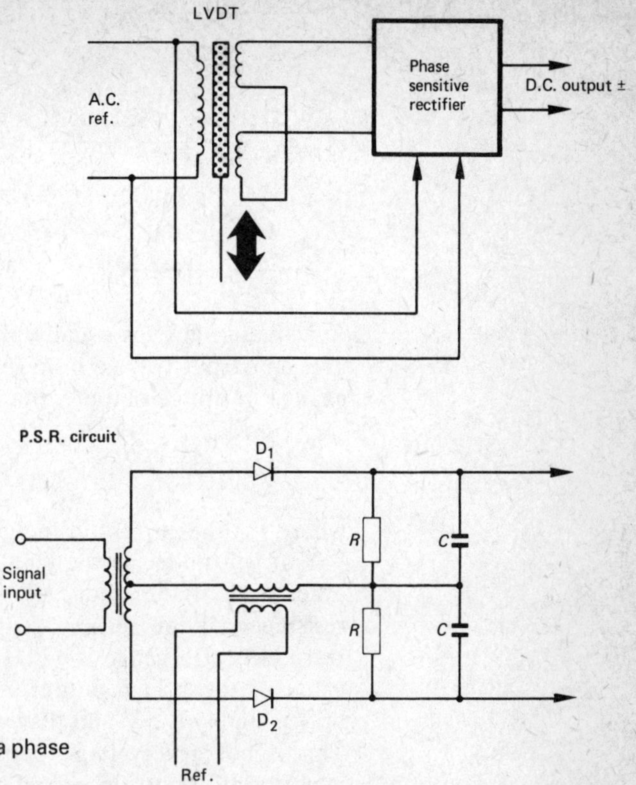

Fig. 4.18b Use of a phase sensitive rectifier

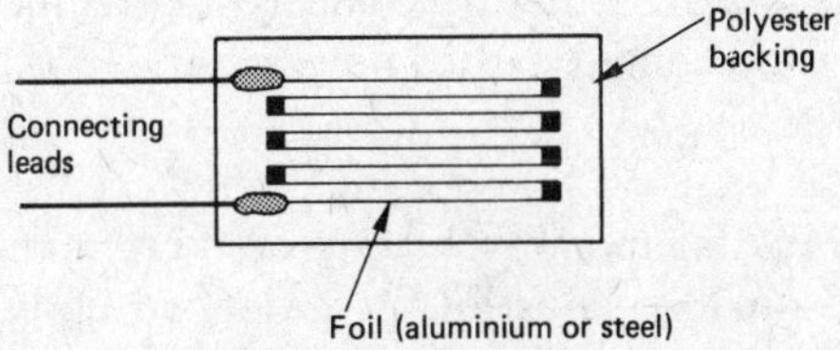

Fig. 4.19 Strain gauge

A change in resistance due to strain can be expressed as

$$\frac{\Delta R}{R} = \frac{\Delta \rho}{\rho} + \frac{\Delta l}{l} - \frac{\Delta a}{a}$$

where $\frac{\Delta l}{l} = \varepsilon$ is the strain, usually expressed in $\mu\text{m}/\text{m}$.

The key parameter for a strain gauge is called the *gauge factor K*:

$$K = \frac{\Delta R/R}{\Delta l/l} = \frac{\Delta R/R}{\varepsilon}$$

A typical value for K, which is a dimensionless quantity, is 2. (Semiconductor gauges are much more sensitive, with gauge factors of 100 or more.)

Fig. 4.20 Strain gauge bridge circuit

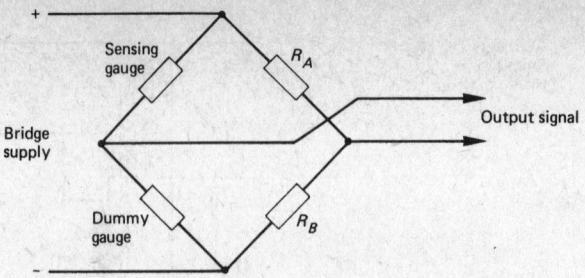

A strain gauge is usually placed in a Wheatstone bridge arrangement to give an output voltage proportional to strain. For the circuit of *fig. 4.20*, the change in bridge output voltage is given by

$$V_o = \frac{R_A R_B}{R_A + R_B} IK\varepsilon$$

where I is the current in the gauge.

You will notice in the circuit that a dummy gauge (unstrained) is in the other arm of the bridge. This provides compensation against changes in resistance of the sensing gauge with temperature. Given that the typical resistance of a gauge is 120 Ω, by fitting 500 Ω resistors for R_A and R_B the value of the bridge output voltage can be calculated. Assuming that the bridge supply is 5 V and that ε (the strain) is 100 μm/m, then $V_o \simeq 160$ mV. This is not a very large output and obviously some form of amplifier is necessary to boost this signal before conversion.

Apart from the direct measurement of strain, the strain gauge is important since it is used to make other transducers such as the load cell, torque meter, accelerometer and some types of flow meter.

4.5 Ultrasonic Devices

Ultrasonic waves (sound waves above 20 kHz) are produced using electro-mechanical resonators. Most of these depend upon the piezo-electric effect and operate at about 40 kHz. A piezo-electric material, such as quartz and certain ceramics, is one that develops a voltage across it if it is mechanically strained in the polarization direction. Conversely, if an electric field is applied along the polarization direction, the material will change its length. Therefore, both output devices, which transmit sound waves at 40 kHz, and sensors of these sound waves can be constructed. An example showing the use of such transducers is given in *fig. 4.21*.

The **transmitter** is a gated oscillator set to run at the resonant frequency of the ultrasonic output transducer. A CMOS 4001B NOR gate IC is used for this circuit which can be switched on and off using logic signals from a micro system. When the logic signal is high (at 1), Tr$_1$ conducts and pin 1 of the IC is taken low to $\emptyset$. The oscillator will then run and ultrasonic waves will be emitted from the transducer.

The **receiver** consists of the ultrasonic sensor, a high gain amplifier, a demodulator and a Schmitt trigger. A burst of 40 kHz sound waves is received

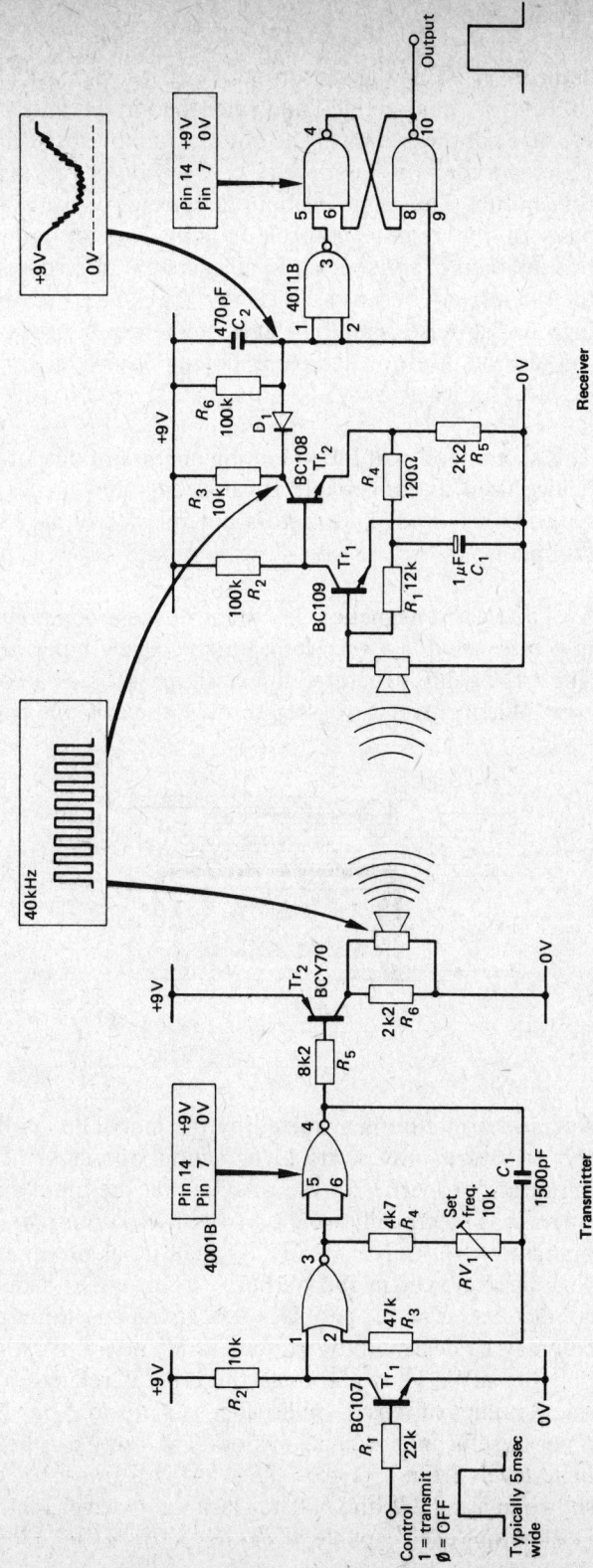

Fig. 4.21 Ultrasonic link

and converted by the sensor into electrical signals. The amplified level at Tr_2 collector is then rectified and smoothed by D_1 and C_2, giving a low signal to the input of the Schmitt. The output from pin 10 of the CMOS 4011B IC goes high and remains high while the sensor is picking up a signal from the transmitter. The range is about 5 metres. A system such as this could be the basis of the remote control of light level or motor speed or simply for transmission of serial data. In all cases the ultrasonic waves are being used as the carrier and, because this is only 40 kHz, the rate of data transfer cannot be high. Input pulses should be at least a few millisecs in width. Ultrasonic waves can also be used for detecting moving objects or for alarm systems.

4.6 Output Devices This section looks at the operation and use of output transducers, the sorts of device used to convert electrical signals into mechanical energy. These are typically solenoids and motors but relays will also be included here in this group.

1 The **electromagnetic relay** is one of those components that have been used in electronics for a very long time and new types are still being developed. The basic relay structure, shown in *fig. 4.22*, is a coil wound on a soft iron core. When current is passed through the coil, the core is magnetised and the

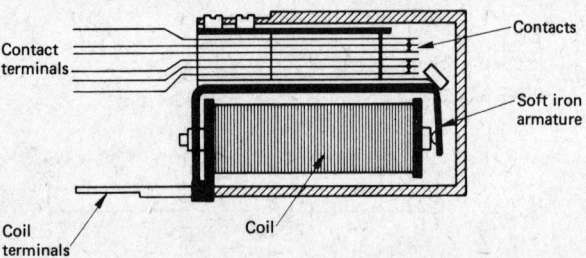

Fig. 4.22 Basic electromagnetic relay

soft iron armature is attracted by the magnetic "pull" towards the core. As the armature moves, its lever action operates the contacts. One of the important properties of the relay is that the input circuit, supplying the coil current, is electrically isolated from the output circuit in which the contacts operate. Also, only a relatively small input power is necessary to control a very large power in the output. For example, a miniature relay may have contacts rated at 3 A and 28 V d.c. giving an output power of 84 W; while its coil may be designed to operate at 6 V and 60 mA which is an input power of only 360 mW. Thus, although the relay is relatively slow, with operate and release times of tens of milliseconds (10 ms to 20 ms is typical), it can provide a very useful interfacing function. The relay can be switched on and off by logic levels from a micro system using a transistor or Darlington switch as shown in *figs 4.23a* and *b*. The Darlington driver IC ULN 2003 contains seven Darlingtons each capable of sinking 500 mA (but not simultaneously since the

Fig. 4.23a Transistor switch to a relay

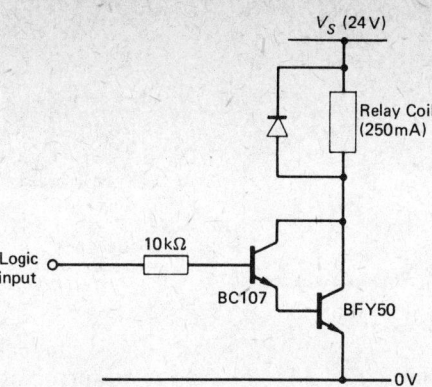

Fig. 4.23b Darlington IC driver to relays

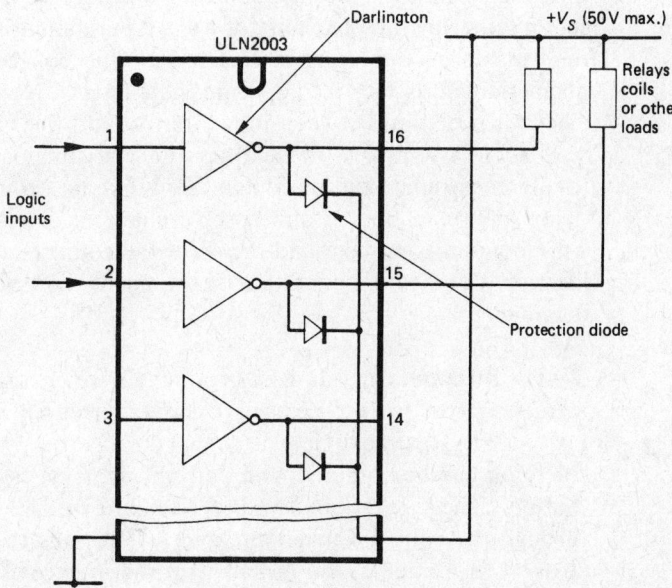

total chip power rating is 1 W at 25°C). Note that in both circuits a protection diode is included to limit the back e.m.f. when the relay is switched off.

The life of a relay will depend upon the type of contact material and the power being switched. Gold-plated contacts are ideal for "dry circuit" switching, i.e. switching at very low levels; while silver-plated contacts are used for higher power applications. Contact life in each case is typically 10^6 operations.

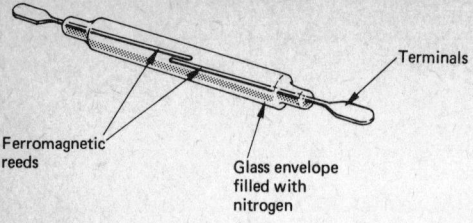

Fig. 4.24 Reed switch

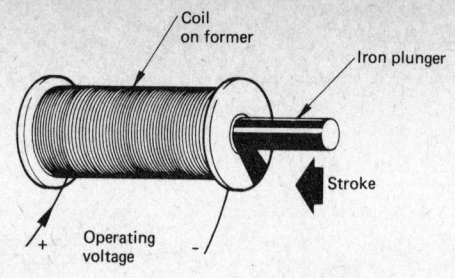

Fig. 4.25a Solenoid construction

2 A **reed relay** (*fig. 4.24*) consists of two ferromagnetic reeds with contacts sealed inside a glass envelope. When in the presence of a magnetic field, the reeds flex together and the contacts make. The operate time is much faster than the conventional relay; about 1 ms is normal. The magnetic field can be set up either by the proximity of a small permanent magnet or by placing the reed inside an operating coil. Then, as the coil current is switched on, a magnetic field is set up to operate the contacts. Note again the high level of electrical isolation between input control and the output. There are several reed relay assemblies, ideal for printed circuit board mounting, which are totally encapsulated in plastic and look like an ordinary dual-in-line IC.

For any relay, apart from those with mercury wetted contacts, the contacts will "bounce" on make and break. If the contacts are used to operate some logic or to indicate a condition to the micro, the switching contacts must be debounced.

3 The **solenoid**, consisting of a coil within which an iron plunger or armature is free to move, is a device used to convert electrical energy into linear motion (*fig. 4.25*). This movement is referred to as the "stroke". General-purpose solenoids can be obtained with pull, thrust, lever (similar to relay), or rotary motion. The operation is quite straightforward; when current is passed through the coil, the magnetic field set up pulls the armature to create the stroke. Movements are usually in the range of 10 mm to 20 mm and manufacturers supply data in the form of a force/stroke graph. Coils can be made to be d.c. or a.c. operated but the d.c types are more efficient. A 12 V d.c. solenoid at 10 W will give a 6 mm stroke with a force of 0.55 kgf. Solenoids, which can be switched on and off in exactly the same way as relay coils, are much more efficient if low duty cycle operation is used, i.e. the device is only pulsed on for short periods. If left on continuously, the coil heats up and efficiency in terms of the force and stroke output falls to a low level.

4 Often in a control system, the output will be required to drive a **motor** of some kind. There are, of course, many different types of motor available, but in this section it will only be possible to review those which are considered the more important. At the same time we can look at various control features such as switching on and off and changing the speed. Apart from these more

Fig. 4.25b Force/stroke graph for a general-purpose solenoid

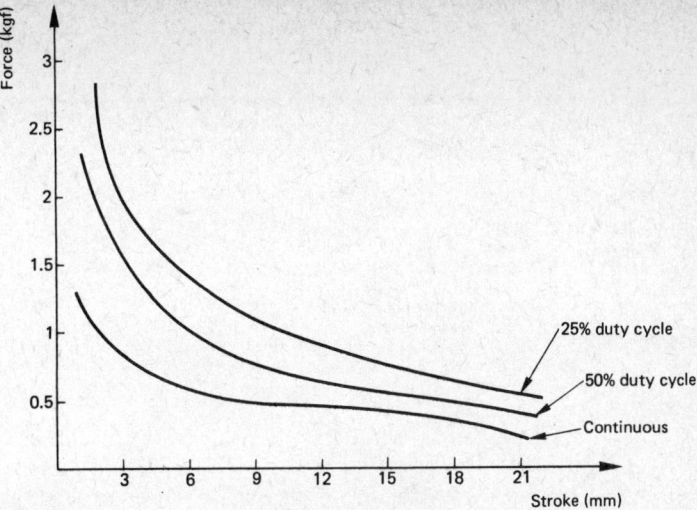

obvious functions, the control of direction, varying position and maintaining output torque may also be significant.

The *motor principle* is that a current-carrying conductor placed in a magnetic field will experience a force tending to pull it into alignment with the field.

$$F = BlI$$

where B = magnetic flux density
l = length of conductor
I = current in the conductor.

In a d.c. motor a commutator is used to reverse the direction of current in the coil so that rotation will be continuous (*fig. 4.26*).

Fig. 4.26 Principle of operation of a d.c. motor

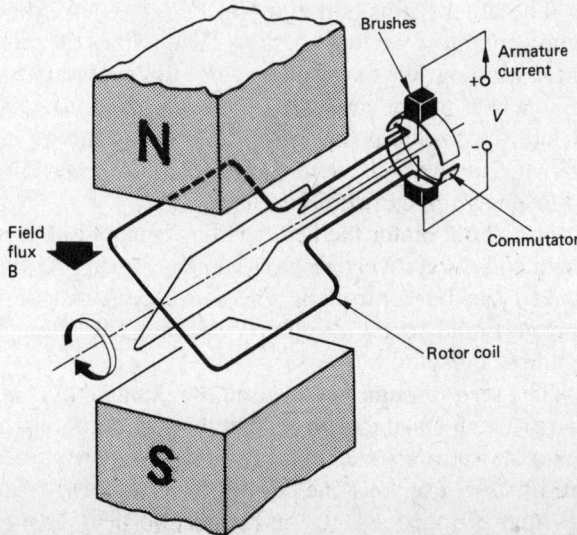

96 Practical Interface Circuits for Micros

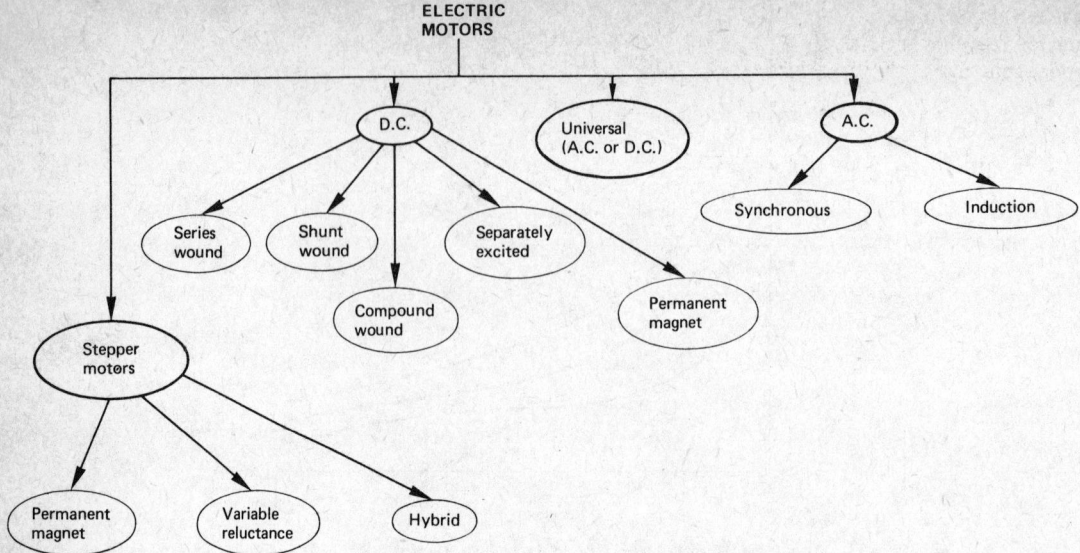

Fig. 4.27 Types of electric motor

A kind of map showing the many types of motor is shown in *fig. 4.27*. For our purposes the most important are the d.c. and stepper motors, and other texts should be consulted for the a.c. motors.

Many of the **smaller d.c. motors** are permanent magnet types, with the magnet providing the necessary field. For these small (fractional horsepower) motors, control of the speed is best achieved by varying the applied d.c. voltage rather than a pulse width method as described earlier. *Fig. 4.28* shows a typical arrangement for speed control of a 12 V 6 W general-purpose permanent-magnet model motor. A four-bit digital word from the output of a micro operates the switches in the CMOS IC. These switches connect binary weighted resistors to a positive supply so that a d.c. voltage is applied via the op-amp and transistor to the motor.

The other forms of d.c. motor (*fig. 4.29*) are the shunt, series, compound, and separately excited versions. The names all refer to the way in which the field winding, the one that provides the necessary magnetic flux, is connected. For a d.c. motor the speed depends on the strength of this magnetic field, while the torque is proportional to the armature current, that is the current flowing in the coils wound on the rotor. Each of the forms therefore have differing characteristics and uses.

In a **shunt motor** the field winding is in parallel with the armature and it can be a coil of relatively high resistance. By varying the field current the motor's speed can be controlled. The torque, which is a medium level, is directly proportional to the armature current. The shunt motor is mainly used for pumps, fans and blowers.

The **series wound motor**, used for example in cranes, lifts and golf carts, has a very high starting torque. The field winding, a coil of a few turns of heavy gauge wire, is connected in series with the armature. The high starting torque results because the same current flows in both armature and field. The torque is then proportional to the product of field strength and armature current.

Sensors and Output Devices

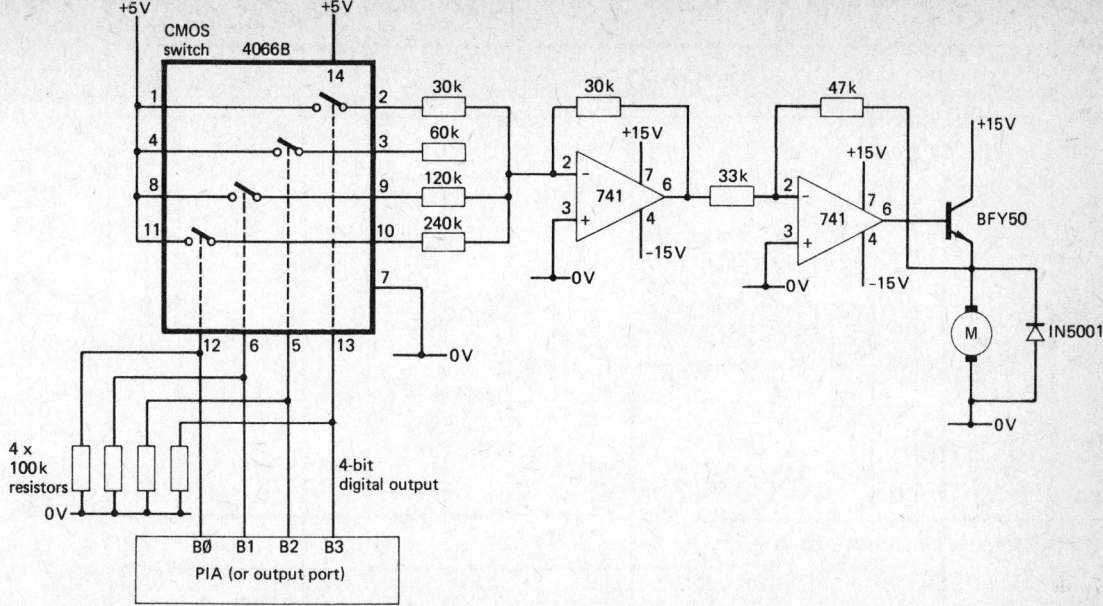

Fig. 4.28 Interface to drive a d.c. motor

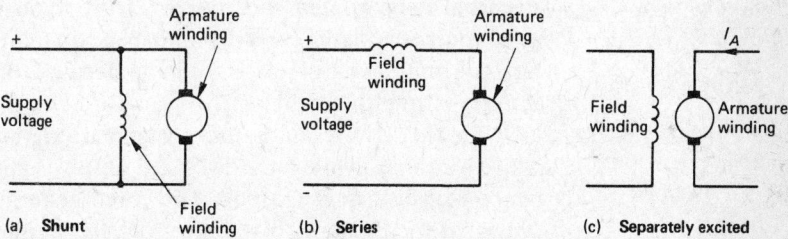

Fig. 4.29 Connection methods for d.c. motors

Note that for a high power rated series motor the load must not be disconnected, otherwise the motor speed would increase to a dangerous level. The small motors used in domestic appliances such as hand drills and food mixers are also examples of the series motor. These are the so-called universal types which can be operated from a d.c. or a.c. supply. Control of speed in a series d.c. motor can be arranged by a pulse width modulated system outlined in *fig. 4.30*. Here the oscillator provides trigger pulses to a monostable circuit and the output pulse from the monostable is used via a drive circuit to switch the motor. For high speed, the monostable pulse width will be set to maximum so that the motor receives a pulse of power that is relatively long in each cycle. For lower speed, the monostable output pulse width will be reduced.

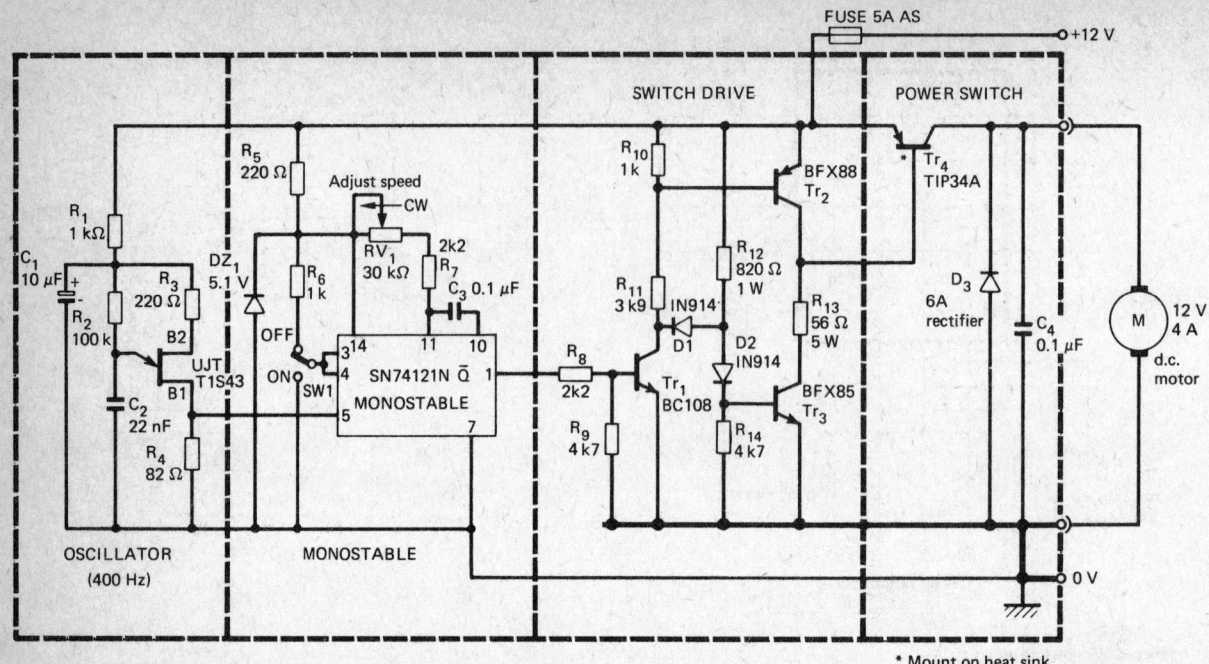

Fig. 4.30 D.C. motor speed control

The **separately excited d.c. motor** is the type most commonly used in position control systems (servos) and speed regulators, especially where the power requirements are relatively low. Because the field and armature are separated, either the field or the armature current can be varied to give control. In the latter, an output voltage from an amplifier is fed to the armature terminals while a fixed voltage supply is applied to the field windings to give a constant field current. The force generated by the motor will be proportional to the applied voltage from the amplifier. Alternatively, if the armature current is held constant and a varying voltage applied to the field winding, the output torque will be proportional to the field voltage. Split field motors are often used in servo systems (position controllers) with the split field driven from the output of a differential amplifier (*fig. 4.31*). When the input to the amplifier is balanced, the two field currents are equal and the motor is at rest. An out-of-balance signal causes the motor to drive in one direction (or the other depending on the polarity of the input signal).

Stepper motors are ideal for use in a microprocessor system for these motors use what is termed direct digital control for movement and position of the rotor (*fig. 4.32*). The general principle is that, when a group of pulses of the correct sequence is applied to the stator windings, the rotor will be advanced through a series of steps. The precise angle of each step depends on the design of the particular motor but is typically 1.8°, 2.5°, 3.75°, 7.5°, 15° or 30°. Stepper motors can be used with a digital system in very simple speed or position controllers under open loop conditions with a high degree of

Fig. 4.31 Using a split field motor in control

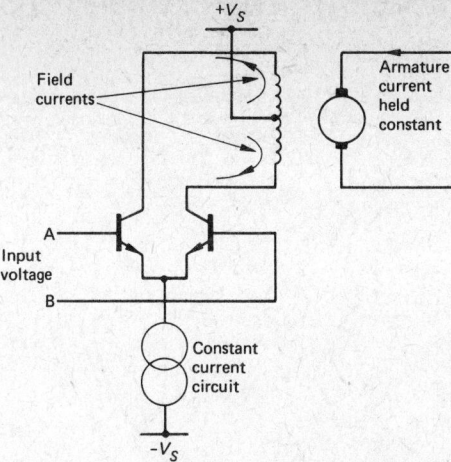

When $V_A = V_B$, the field currents are equal and opposite so the motor is at rest.

Fig. 4.32 General view of a stepper motor

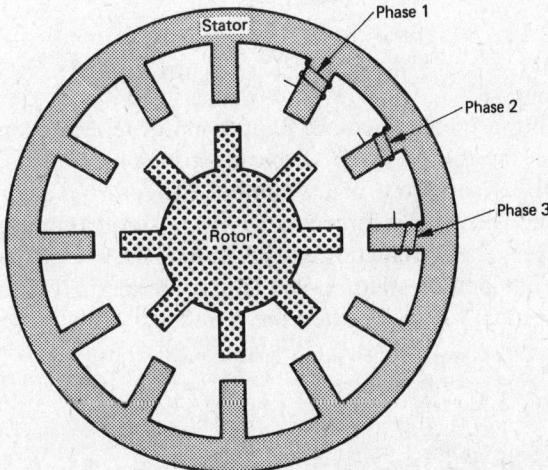

precision and accuracy. This follows because the number of pulses sent out by the digital controller can be easily counted.

There are three main versions of stepper motors. The **permanent magnet type** has a permanent magnet rotor with a large number of poles. The stator construction is such that, when pulses are applied to the stator windings, the poles, which have the same number as the rotor poles, reverse. The rotor is then attracted to align itself so that each of its S-poles is mid-way between a pair of N-poles on the stator, and each rotor N-pole is mid-way between a pair of S-poles on the stator. Speed is controlled by the rate of pulses applied to the stator and position is controlled by a fixed number of pulses. Because there is a force of attraction always in existence between the rotor poles and stator poles, this type of stepper motor has what is called detent torque. In other words, a torque must be applied to displace the rotor from its rest position even when the motor is unpowered. This useful feature is not possessed by other types.

Fig. 4.33 Four-phase variable-reluctance stepper motor

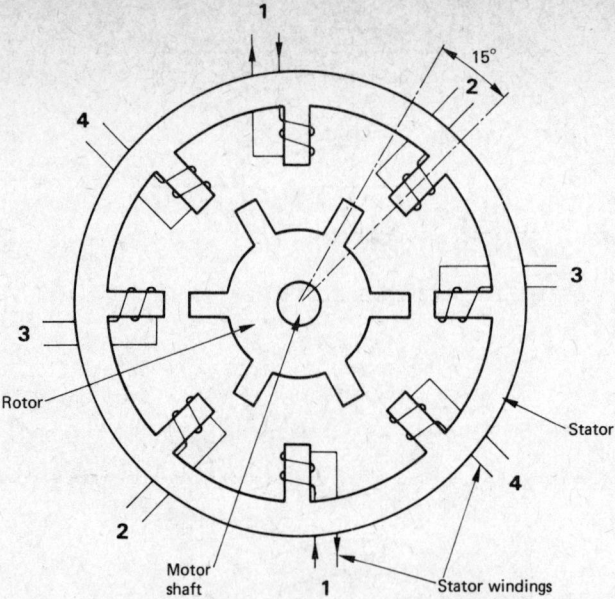

The **variable reluctance** stepper motor requires a multi-phase drive signal to cause the rotor to step. Typically a sequence of four pulses is required. The rotor, of soft iron, has a number of teeth unequal to the number of stator poles. The rotor is forced to change position as its aligns itself with the path of least magnetic reluctance. The step angle, set by the design, is controlled by the number of stator poles and rotor teeth. For example, if a motor has $S_P = 16$ and $R_T = 12$, then the number of steps per revolution is

$$N = \frac{S_P R_T}{S_P - R_T} = 48$$

∴ Step angle = $360°/48 = 7.5°$

The variable reluctance stepper motor is capable of operation at high stepping rates. The general principle of its operation can be seen from *fig. 4.33* for a motor with 8 stator poles and 6 rotor teeth (step angle = 15°). When phase 1 of the stator is energised, two of the rotor teeth line up with the stator while the others are 15° out of alignment. Energising phase 2 causes the rotor step 15° to align two of the other teeth and so on. If the stator coils are energised sequentially 1–2–3–4, the rotor will move in 15° steps counter-clockwise.

The third type is the **hybrid stepper** motor (*fig. 4.34*), which as its name implies combines construction features of both the PM and VR types. The stator has typically eight salient poles which are activated by a two-phase winding. The rotor is an axially magnetised cylindrical magnet. For the hybrid, the step angle is small, between 0.9° to 5°, with 1.8° step being the most common. This gives 200 steps per revolution.

There are some general terms used in specifying stepper motors as follows (*fig. 4.35*):

Fig. 4.34 Hybrid stepper motor

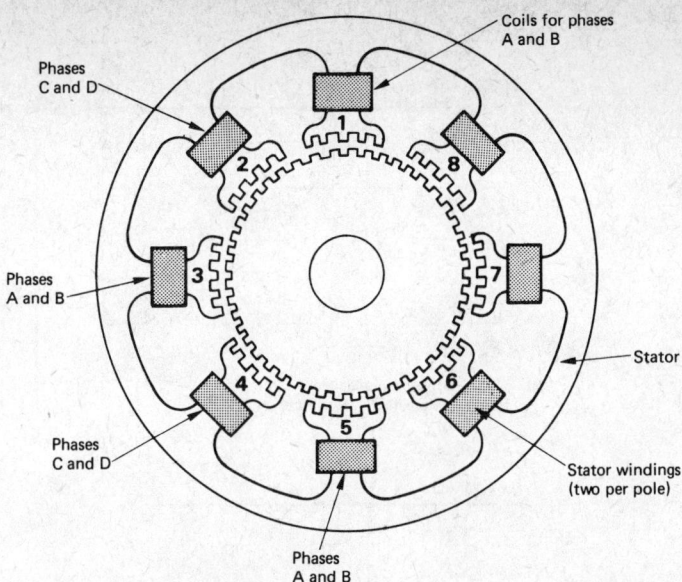

Holding torque The maximum torque that can be applied to a powered motor without moving it from its rest position (note this is not the same as detent torque).

Pull-in torque The maximum torque against which a motor will start, for a given pulse rate, and reach synchronism without losing a step.

Pull-out torque The maximum torque that can be applied to a motor, running at a given stepping rate, without losing synchronism.

Pull-in rate The maximum switching rate at which a loaded motor can start without losing a step.

Pull-out rate The switching rate at which a loaded motor will remain in synchronism as the switching rate is reduced.

Slew-range The range of switching rates between pull-in and pull-out within which the motor runs in synchronism but cannot start up (or reverse).

Fig. 4.35 Stepper motor characteristics

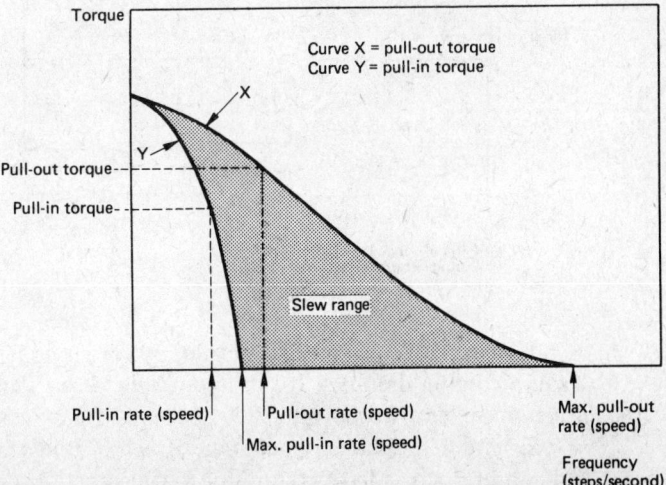

Fig. 4.36 Using the SAA 1027 to drive a four-phase stepper motor

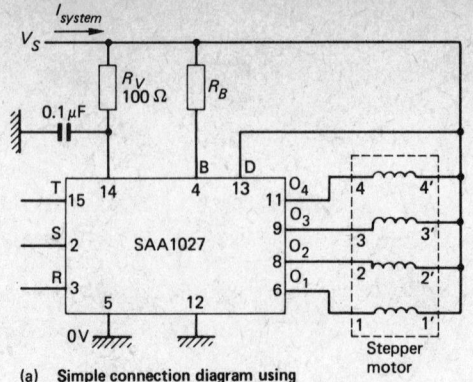

(a) Simple connection diagram using internal diode dissipation circuit

T is the **Trigger** input. Pulses on this pin set the stepping rate and therefore the angular position of the shaft.

S is the **Set** input. This allows the output switching sequence to be set to a predetermined state by applying a low input. Effective only if the T input is high. If not required, connect to $+V_S$.

R is the direction of **Rotation** input.
High = counterclockwise
Low = clockwise

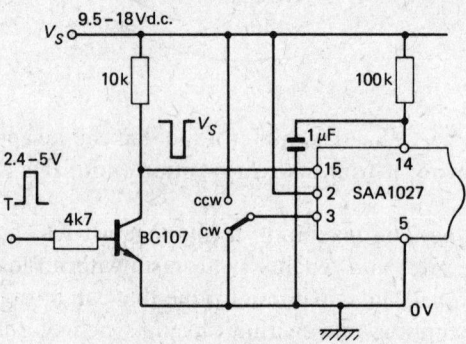

(b) Typical interface to TTL-type signals with connections for T, S and R shown

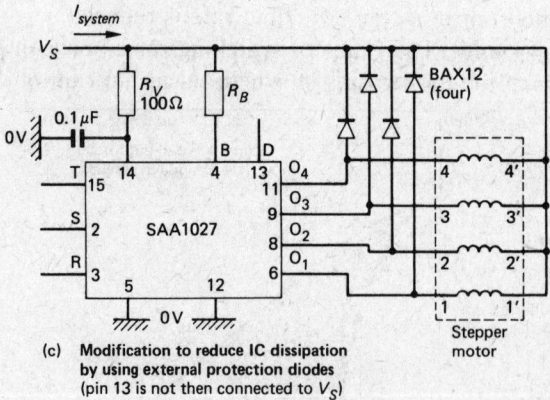

(c) Modification to reduce IC dissipation by using external protection diodes (pin 13 is not then connected to V_S)

Specialised ICs have been developed for stepper motor drive circuits, one example being the SAA 1027 shown in *fig. 4.36*. This IC, intended for driving four-phase stepper motors (such as the ID04), consists of a three-input stage, a logic circuit, and an output stage. Each output can supply up to 350 mA of drive and is provided with an internal protection diode.

5 Conversion

5.1 Data Conversion

An understanding of the methods used to convert the analog inputs from sensors into digital signals, and digital output commands from a micro back into analog form to drive output devices, is crucial in the design and construction of digital control systems. To begin with, there are several specific parameters used to express the performance of convertors. One of these, for example, is the **conversion time**: in other words how fast the conversion can be carried out. A-to-D convertors range from the fairly slow, with conversion times of many milliseconds, to the very fast types that can complete a conversion in a few nanoseconds. Obviously the latter, called flash convertors, are more costly and will not be used unless speed is the prime consideration. Other important aspects concerning convertors are :

1 An appreciation of the various errors involved in the conversion process.
2 Ensuring that the number of bits used for the conversion is appropriate to the input/output requirements.

The number of bits determines the resolution. But it should be noted that high resolution does not necessarily imply high accuracy, for a convertor with a large number of bits may still have errors. Table 5.1 shows the value of resolution obtained for the number of bits used in a conversion. For an example consider a 3-bit convertor (A-to-D *or* D-to-A). The digital word of 3 bits can range from ØØØ up to 111 which will give 8 possible analog states or levels and a resolution of 12.5%. In other words, if the full-scale value of the analog signal is 8.75 V, each discrete step of the analog signal involved in the conversion will be separated from the next by 1.25 V (*fig. 5.1*). This can be expressed as a table:

Digital signal			Analog signal
MSB b2	b1	LSB bØ	Volts
Ø	Ø	Ø	0
Ø	Ø	1	1.25
Ø	1	Ø	2.5
Ø	1	1	3.75
1	Ø	Ø	5
1	Ø	1	6.25
1	1	Ø	7.5
1	1	1	8.75

Table 5.1

Word length in bits (n)	Maximum number of possible combinations (2^n)	Decimal digits	Equivalent accuracy or resolution of least significant bit	
			percent	ppm
1	2	1	50.	500 000.
2	4	1	25.	250 000.
3	8	1	12.5	125 000.
4	16	2	6.25	62 500.
5	32	2	3.125	31 250.
6	64	2	1.562 5	15 625.
7	128	3	0.781 25	7 812.5
8	256	3	0.390 625	3 906.25
9	512	3	0.195 313	1 953.13
10	1 024	4	0.097 656	976.56
11	2 048	4	0.048 828	488.28
12	4 096	4	0.024 414	244.14
13	8 192	4	0.012 207	122.07
14	16 384	5	0.006 104	61.04
15	32 768	5	0.003 052	30.52
16	65 536	5	0.001 526	15.26
17	131 072	6	0.000 763	7.63
18	262 144	6	0.000 381	3.81
19	524 288	6	0.000 191	1.91
20	1 048 576	7	0.000 095	0.95
21	2 097 152	7	0.000 048	0.48
22	4 194 304	7	0.000 024	0.24
23	8 388 608	7	0.000 012	0.12
24	16 777 216	8	0.000 006	0.06

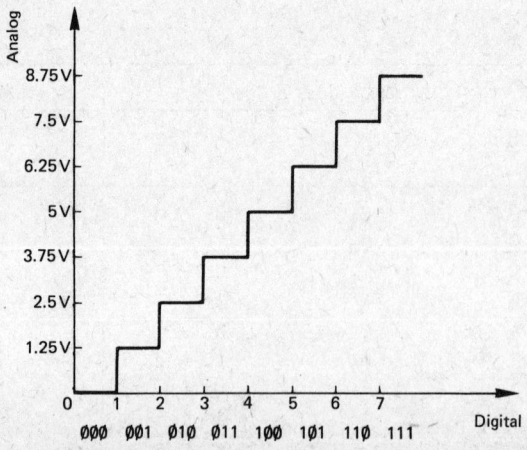

Fig. 5.1 Transfer characteristic of a convertor

On the other hand, an 8-bit convertor will have 256 possible states ($2^8 = 256$) and the resolution of the LSB, for a full-scale output of 10 V, is a mere 39 mV. A large number of bits must be used when a fairly faithful conversion between analog and digital (and vice versa) is required.

Since many A-to-D convertor circuits require a D-to-A convertor as one of the main elements we shall look at the DAC first.

5.2 Digital-to-Analog Conversion

The principle of a DAC is that, when a digital input is applied, the analog output, which may be a current or voltage, takes a value according to the weight of the digital input. This is usually called a multiplying digital-to-analog convertor because it gives a way of obtaining the continuous multiplication of the digital input value; the product being represented by the varying analog output.

If a unipolar voltage output and normal binary coding for the digital input are assumed, then the transfer function of a DAC can be written using an equation. For the example in *fig. 5.1* the equation will be

$$V_{out} = V_{ref} \cdot \left(\frac{B_2}{2} + \frac{B_1}{4} + \frac{B_0}{8} \right)$$

where B_2 is the most significant bit (MSB) and B_0 is the least significant bit (LSB). A general equation takes the form

$$V_{out} = V_{ref} \cdot \left(\frac{B_{(n-1)}}{2} + \frac{B_{(n-2)}}{4} + \frac{B_{(n-3)}}{8} + \ldots + \frac{B_0}{2^n} \right)$$

where n = the number of bits used in the convertor.

The bits can of course each take a value of logic 1 or logic 0. Thus in the 3-bit DAC, if the digital input is 101 and $V_{ref} = 10$ V,

$$V_{out} = 10\left(\frac{1}{2} + \frac{0}{4} + \frac{1}{8}\right) = 6.25 \text{ V}$$

The maximum output from a DAC is called the *full-scale output* (V_{FSO}) and this occurs when *all* the bits in the digital input are high at logic 1. In the case of the 3-bit convertor,

$$V_{FSO} = \tfrac{7}{8} V_{ref} = 8.75 \text{ V}$$

But the V_{FSO} for a 6-bit convertor which has $V_{ref} = 10$ V will be

$$V_{out} = 10\left(\frac{1}{2} + \frac{1}{4} + \frac{1}{8} + \frac{1}{16} + \frac{1}{32} + \frac{1}{64}\right)$$

$$= 9.843\,75 \text{ V}$$

For the 6-bit DAC, each step in the analog output is 156.25 mV.

We now look at the various errors that can occur in a DAC. **Errors**, however small, will arise because of the mismatch of resistors and because the electronic switches inside the DAC will not have zero "on" resistance. As shown in *fig. 5.2*, as the digital input code to the DAC is increased by one

Fig. 5.2 Principle of digital-to-analog conversion

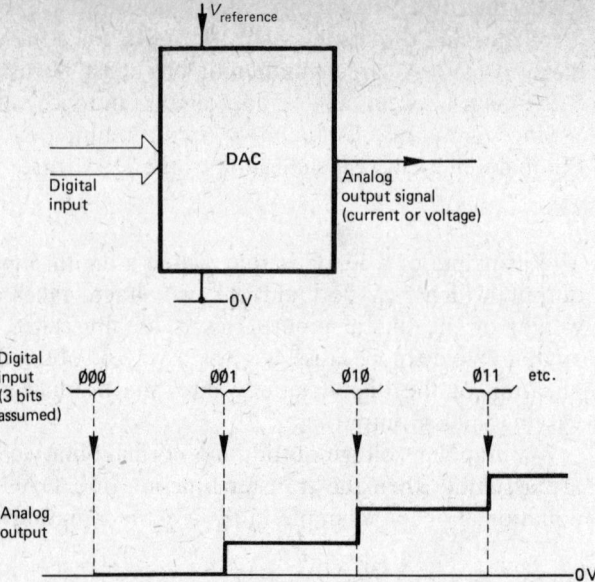

LSB at a time, the analog output should also increase uniformly giving a "staircase" type waveform. As long as the output does increase in this manner, the DAC is said to be monotonic. Errors can, of course, force the output to be non-monotonic. This could happen if the working temperature increased beyond the specification limits. *Figure 5.3* illustrates a non-monotonic DAC, where the analog output for the digital input code 1Ø1 falls instead of rising. This example is an obvious case of a nonlinear transfer function.

For a DAC, **nonlinearity** is defined as the maximum amount by which any point on the transfer characteristic deviates from the ideal straight line. Nonlinearity is usually expressed as a fraction of an LSB. Typically, the value is $\pm\frac{1}{2}$LSB and a nonlinearity of this size will mean that the DAC remains monotonic. Another type of nonlinearity that can occur, called **differential nonlinearity**, is an error between the value or height of any one of the analog step outputs compared to the others. Using the 3-bit DAC as an example, each step should be 1.25 V (for $V_{ref} = 10$ V); but if one step is 1.875 V then a differential nonlinearity of $+\frac{1}{2}$LSB exists. Typically, a DAC will have a specified differential nonlinearity of $\pm\frac{1}{2}$LSB.

Suppose, however, that when tested by the manufacturer a DAC is found to have a differential nonlinearity of say $-1\frac{1}{2}$ LSB, it will then be non-monotonic. But the device could still be offered for sale with a good price advantage by stating that it has a reduced resolution of 7 bits instead of 8. To do this the LSB digital input will be permanently connected to logic Ø.

Apart from nonlinearity two other errors occur:

1 Offset error: the small value of output voltage (or current) which appears at the output when all the DAC digital inputs are at logic Ø (*fig. 5.4*). This can be readily trimmed out by a simple circuit modification (*fig. 5.5*).

Fig. 5.3 A non-monotonic response

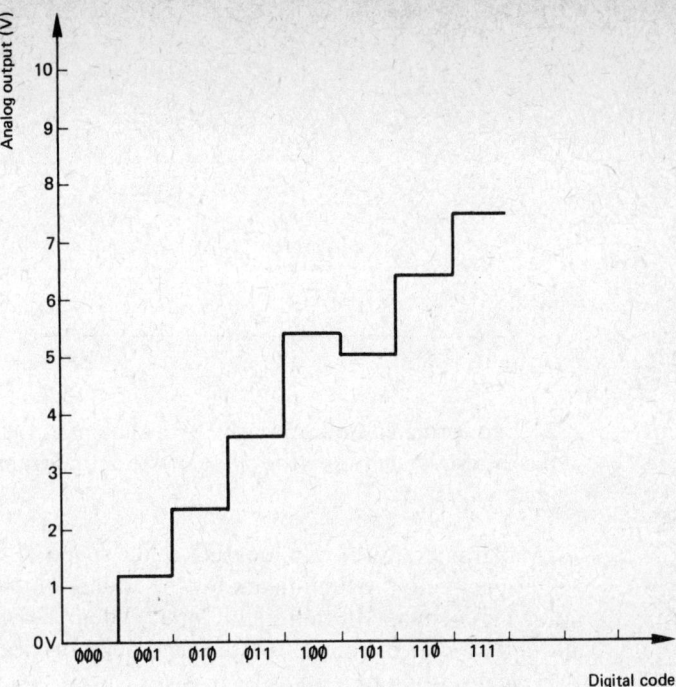

Fig. 5.4 Offset error in a DAC

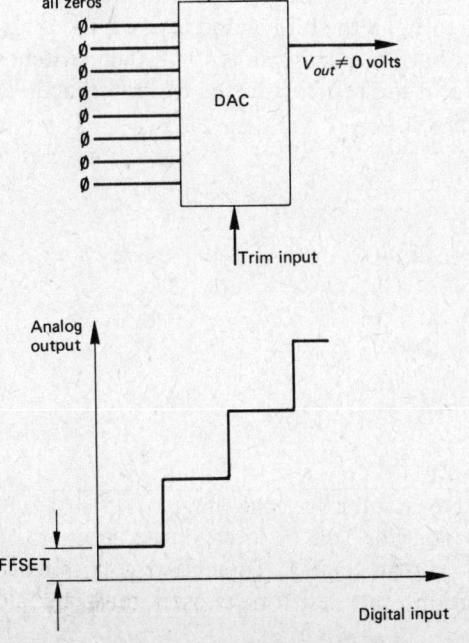

Fig. 5.5 Removal of offset error

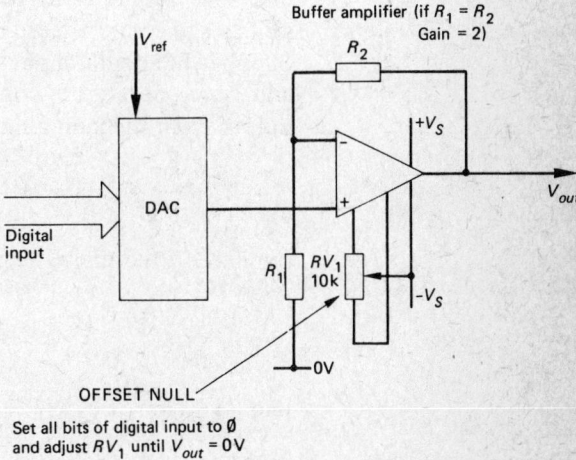

Fig. 5.6 Gain error in a DAC

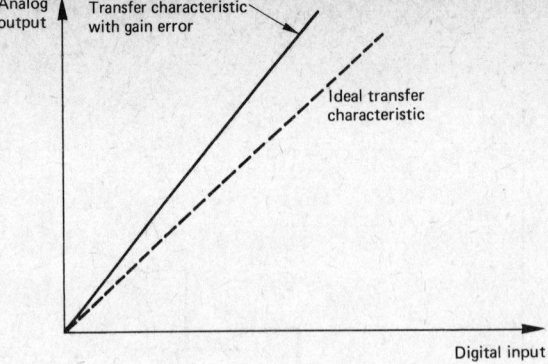

2 Gain error: caused primarily by a change in the reference voltage; this is the difference between the slope of the actual transfer characteristic and the ideal (*fig. 5.6*).

The parameter used as a measure of the speed of a DAC is *settling time*: the value specifying the time taken for the analog output to settle within $\pm\frac{1}{2}$LSB following a change in the digital input code. The worst-case change is when all the bits switch from 1 to $\emptyset$ (1111 to $\emptyset\emptyset\emptyset\emptyset$ or vice versa for a 4-bit DAC). A typical settling time is 1 μs.

5.3 Practical DACs

The simplest method of building a DAC is to use a **weighted-resistor network**, a summing amplifier, and a set of electronic switches (*fig. 5.7*). Each of the resistors has to be weighted in value in a binary sequence, i.e. R, 2R, 4R, 8R, etc. The digital word to be converted is used to operate the electronic switches to connect these resistors to V_{ref} if the bit is 1 and to 0 V if the bit is $\emptyset$. Suppose the digital input for the 4-bit DAC example is 1$\emptyset$1$\emptyset$; then switches 1 and 3 are operated to connect two of the resistors to the 5 V reference. The output from the summing op-amp will be

$$V_o = \frac{R_f}{R} \cdot V_{ref} (1 + \tfrac{1}{4}) = 3.125 \text{ V}$$

Similarly if the digital input changes to $\emptyset$11$\emptyset$, then

$$V_o = \frac{R_f}{R} \cdot V_{ref} (\tfrac{1}{2} + \tfrac{1}{4}) = 1.875 \text{ V}$$

and $V_{FSO} = \frac{R_f}{R} \cdot V_{ref} (1 + \tfrac{1}{2} + \tfrac{1}{4} + \tfrac{1}{8}) = 4.6875 \text{ V}$

V_{FSO} occurs when the digital input is 1111.

The only problem with this simple circuit is that the range of resistor values required for a high resolution convertor will be quite large. For a 12-bit convertor the resistor range is more than 2000:1. To achieve good linearity, accuracy, and monotonic operation, the resistors chosen must be close

Fig. 5.7 Four-bit DAC using a binary weighted network (shown with the input set to 101∅)

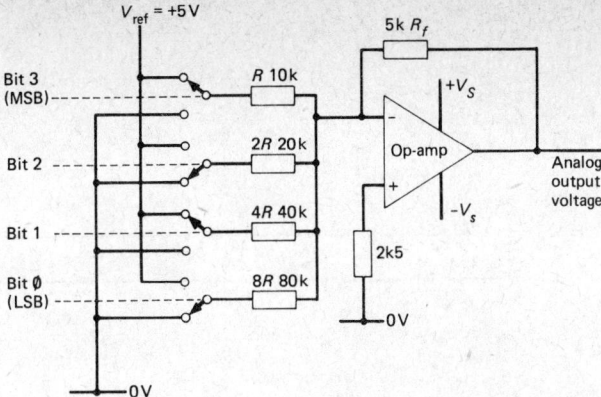

Fig. 5.8 R-2R ladder (3-bit DAC shown)

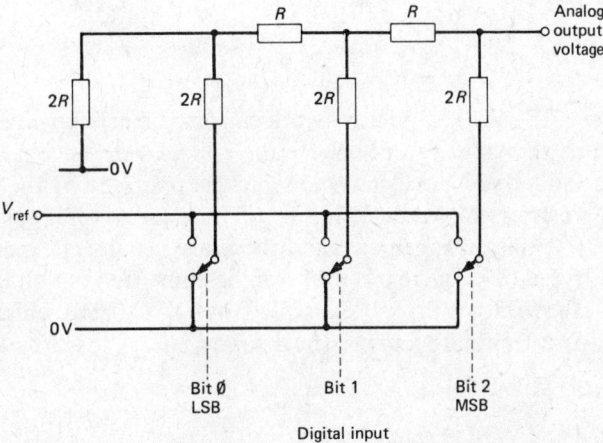

tolerance and must all track together with temperature. This becomes very difficult for DACs of more than a few bits. Therefore, although this simple circuit is useful for low resolution convertors, another method of conversion is preferred.

The most commonly used system for DACs is based on the **R-2R ladder network** shown in *fig. 5.8*. The output voltage is generated by switching sections of the ladder to either V_{ref} or 0 V corresponding to a 1 or a ∅ of the digital input. The switches are electronic and are usually incorporated in the DAC IC.

There are several advantages of the R-2R ladder compared to the weighted-resistor network:

1 Only two values of resistors are used.
2 It can easily be extended to as many bits as desired.
3 The absolute value of the resistor is not important, only the ratio needs to be exact.
4 The resistor network can be fairly readily manufactured as a film network or in monolithic form. In this way the temperature characteristics of the resistors will all be very similar.

Fig. 5.9 R-2R current-output DAC (3-bits only shown)

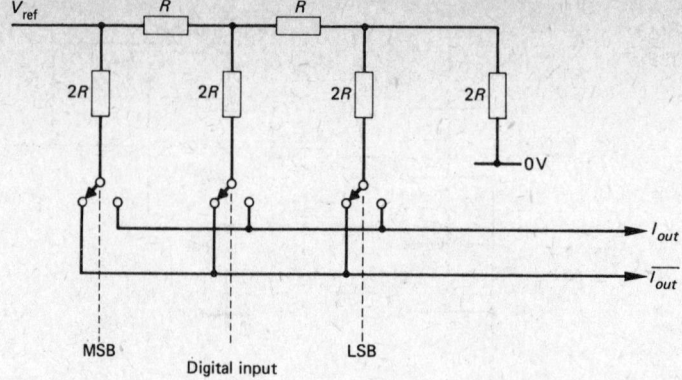

The R-2R ladder network can also be used to give an analog output current as shown in *fig. 5.9* where the voltage reference is applied to one end of the ladder and the electronic switches, operated by the digital input, connect the resistors to either I_{out} or $\overline{I_{out}}$. There are several commercial DAC ICs which use this arrangement, and these are usually cheaper than the voltage output types. The current output can be converted to voltage quite easily, and this is shown for a typical example (the DAC 0800) which is discussed later. Four DAC circuit examples now follow.

1 ZN434 (Ferranti) (fig. 5.10)
This is a 4-bit DAC IC consisting of bipolar switches, an R-2R ladder, and a built-in amplifier and attenuator which provides a reference voltage of nominally $\frac{1}{2}V_{CC}$. The linearity is $\pm\frac{1}{4}$LSB; the chip is TTL and CMOS compatible and has a fast settling time (300 nsec for a digital input change from ØØØØ to 1111 or vice versa). The output resistance is nominally 2.5 kΩ and, therefore, in cases where a relatively heavy load is used (less than say 50 kΩ), a buffer amplifier must be used. This has the added advantage that an analog output of greater than $\frac{1}{2}V_{CC}$ is possible, and also that the small zero offset can be trimmed out, thus ensuring that the analog output for a digital input of ØØØØ is exactly zero volts. A 531 op-amp is used in a non-inverting configuration with offset null provided by RV_1 and gain adjust by RV_2. The gain is given by

$$A_v = \frac{R_1 + R_2}{R_1}$$

and R_1 in parallel with R_2 should equal the output resistance of the DAC in order to reduce drift with temperature. Therefore with $R_1 = R_2 = 5$ kΩ, the gain is 2 and the analog output will have a full-scale value of nearly 5 V. The amplifier circuit gives unipolar output, i.e. the analog output increases from

Fig. 5.10 The ZN434 4-bit DAC with buffer op-amp

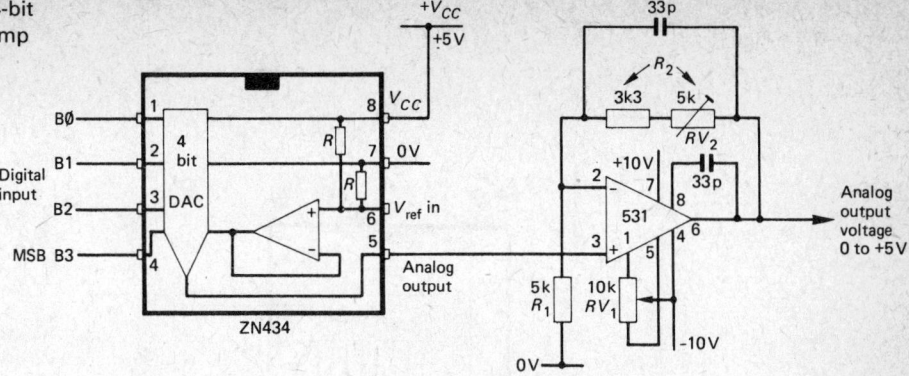

Fig. 5.11 Modification to op-amp to give bipolar output voltage (± 5 V)

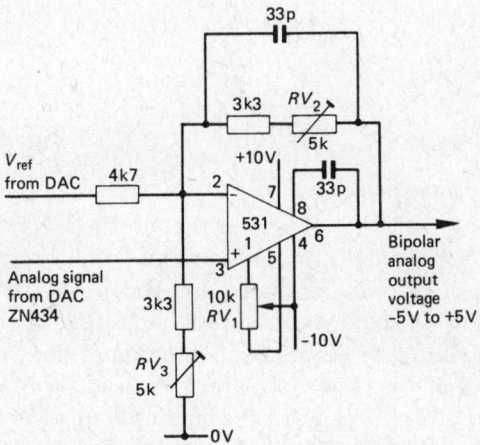

zero to +5 V as the digital input is increased. For bipolar operation, giving a ±5 V output, the amplifier is modified by connecting a third resistor (*fig. 5.11*) to the inverting input from the V_{ref} of the DAC. For all input codes where the MSB is ∅, the output voltage will be negative but, when the digital input has the MSB = 1 (from 1∅∅∅ through to 1111), the analog output becomes positive. For this type of DAC the input code is referred to as **offset binary**. Using a DAC in this way to give bipolar outputs enables true alternating signals to be generated at the analog output. These can be at very low frequencies determined by the rate at which digital words are output to the DAC.

Fig. 5.12 The ZN429 8-bit DAC and buffer amplifier

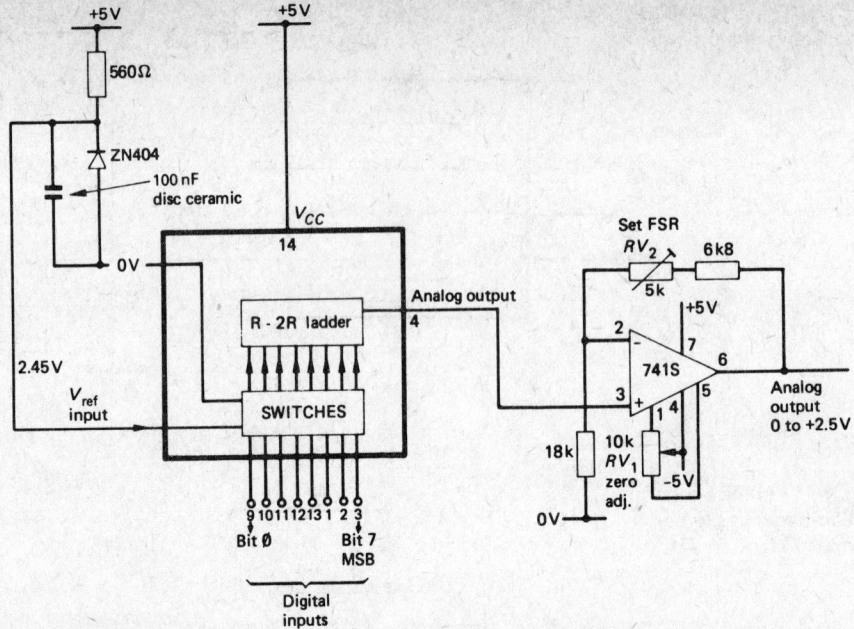

2 ZN429 (*fig. 5.12*)

This is an 8-bit DAC of similar type to the ZN434 except that an external reference supply (V_{ref}) is required. This voltage reference should be between 2.0 V and 3.0 V, well regulated and decoupled with capacitors as shown to reduce noise. The manufacturers state that the slope resistance of the reference should be less than $2\,\Omega$, therefore the circuit shown with *fig. 5.12* would be suitable. The ZN429 E-8, with an 8-bit accuracy, has a maximum nonlinearity of $\pm\,0.5\text{LSB}$ and a settling time of typically $2\,\mu\text{sec}$. The output buffer matched to the output resistance of the DAC ($10\,\text{k}\Omega$) uses a 741S op-amp to give a unipolar output. As for the previous example the circuit can be simply modified to give a bipolar output.

3 ZN428 (*fig. 5.13*)

For micro systems it is often useful to have a data latch built into the DAC since this allows the DAC to be updated direct from the data bus. The ZN428, an 8-bit DAC, has this extra facility. A control input (on pin 4) called **ENABLE** allows data to be set or held. The action is shown in the table.

ENABLE	Result
Low ∅	Latch is transparent
High 1	Data held

In an application example (*fig. 5.14*), the ZN428 data inputs are directly connected to the microprocessor's data bus and the ENABLE signal can be obtained from address decoder logic. When the DAC address is called up, and the VMA (∅2 of the clock on the M6800) and the WRITE signals are

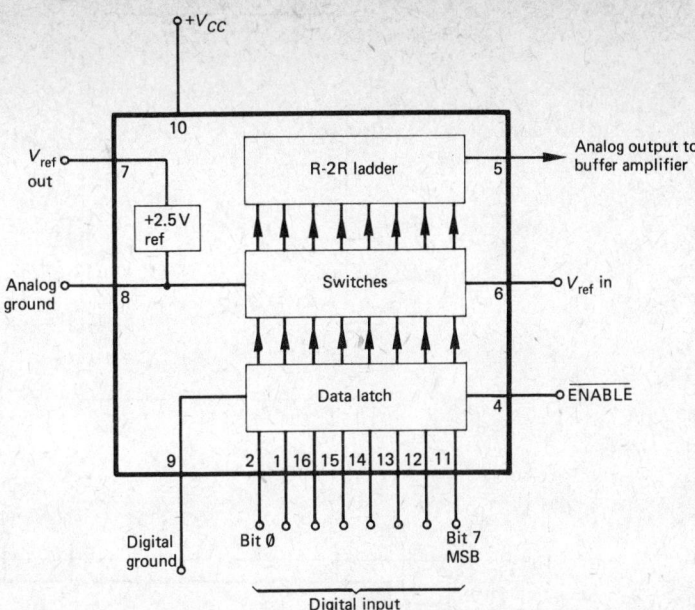

Fig. 5.13 The ZN428 8-bit DAC

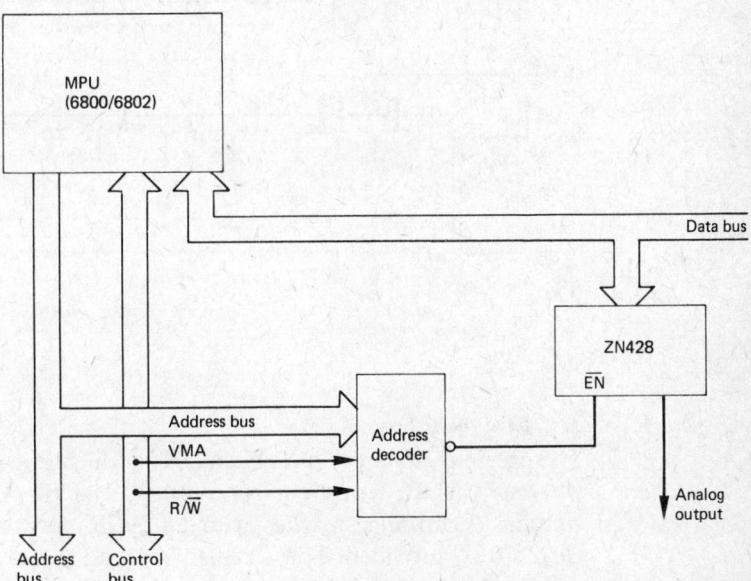

Fig. 5.14 Interfacing the ZN428 directly with a 6800/6802 system

present, then a negative going pulse is applied to the ENABLE pin of the ZN428 to allow the word on the data bus to be transferred into the latches inside the DAC. This data will be held in the DAC until the DAC is again addressed. A system like this is termed **Memory Mapped I/O** and has obvious advantages when several DACs are required.

There are other DACs available that can be used in memory mapped I/O systems, some of which have the address decode logic already built in. The 7542 12-bit DAC is typical.

Fig. 5.15 The DAC 0800

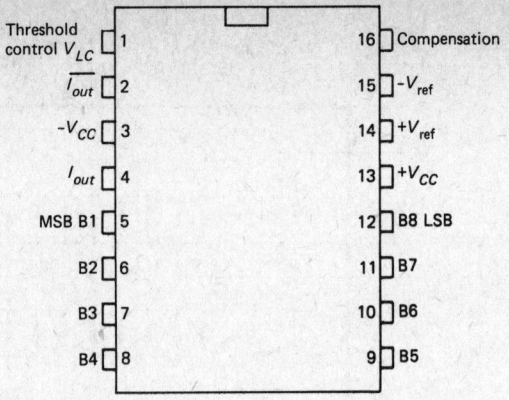

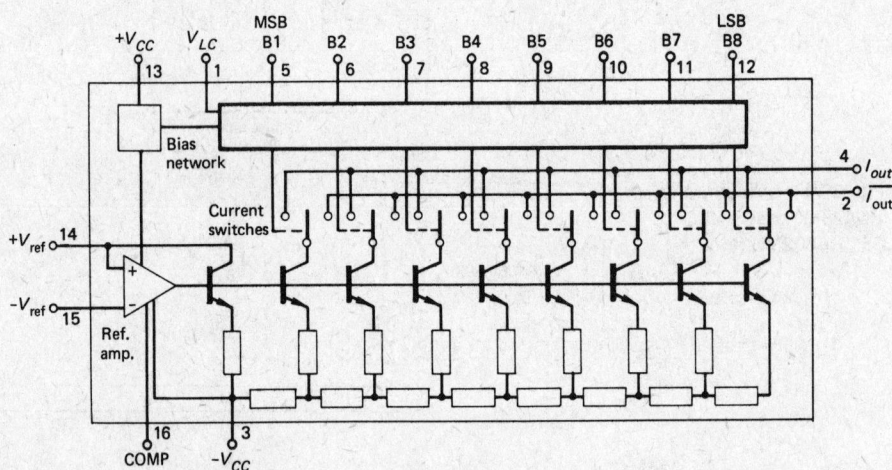

(a) Pin-out and architecture

4 DAC 0800 (fig. 5.15)

The previous examples have been DACs which give an analog voltage output, whereas this IC gives current output. The block diagram shows that it contains electronic switches with an R-2R network, and that two output currents are provided. These currents, I_o and $\overline{I}_o$, are complementary; in other words when $I_o = 1.992\,\text{mA}$, $\overline{I}_o = 0.000\,\text{mA}$, but when $I_o = 0.000\,\text{mA}$, $\overline{I}_o = 1.992\,\text{mA}$. The relationships between these currents and other circuit variables are as follows:

$$I_o + \overline{I}_o = I_{FS}$$

where I_{FS} = full-scale output current in mA

$$I_{FS} = \frac{V_{\text{ref}}}{R_{\text{ref}}} \times \frac{255}{256}\,\text{mA}$$

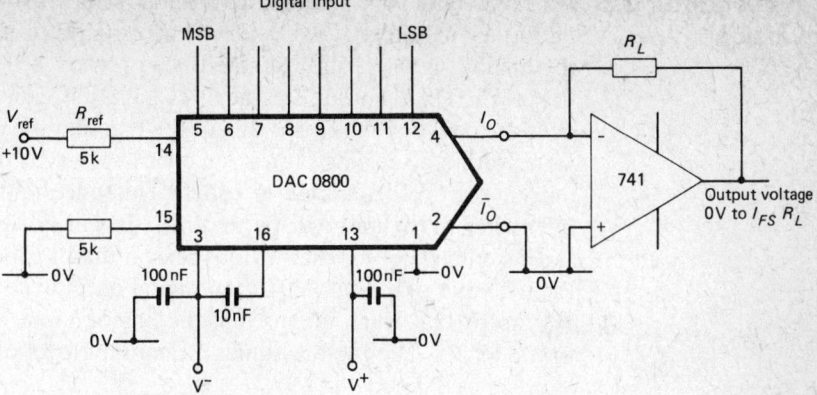

(b) Obtaining a positive output voltage

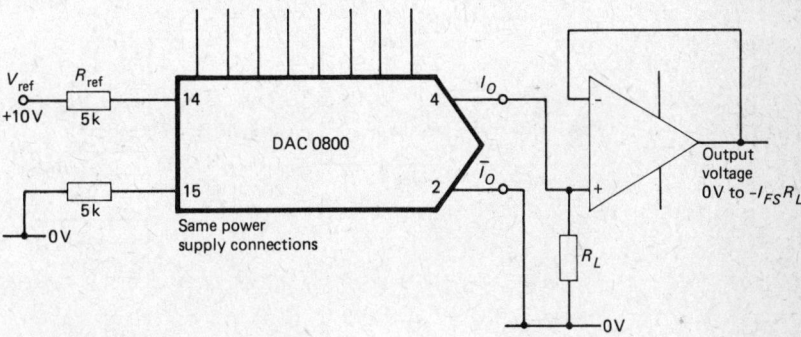

(c) Obtaining a negative output voltage

Normally, I_{FS} is set to equal 1.992 mA, with the reference current at 2 mA. As for any DAC which has not got a built-in reference supply, the external reference must be well regulated and decoupled to ground with a bypass capacitor. Two simple circuits for converting the output current into positive and negative voltage are shown. Both use op-amps to give a low impedance output and the output swing can be adjusted by changing the value of R_L. Note that an offset nulling circuit should be used and this should be adjusted to give zero output voltage when all bits of the digital input are at logic ∅.

The DAC 0800 is designed to operate with TTL-type inputs but by changing the voltage level applied to pin 1 (threshold control V_{LC}) interfacing with any other types of logic is possible.

5.4 Analog-to-Digital Conversion

An ADC takes the analog input signal, samples it, and then produces as its output a coded digital word which corresponds to the level of that portion of the analog signal being examined. The process (*fig. 5.16*) is the opposite to that of the DAC but, unlike the DAC, an ADC will contain some uncertainty over the conversion. This is due to the fact that the analog input can take any value within a set range for it is a continuous type of signal, while the digital output is a fixed number of codes. This uncertainty for an ADC is called **quantising error** and will be $\pm \frac{1}{2}$LSB. It will not be quoted on a manufacturer's data sheet for the obvious reason that it is bound to be present. Take the example of a 3-bit ADC: the digital output can have a coded value from 000 up to 111 which means that the analog input is split up or "quantised" into 8 levels. The table assumes a quantum level of 350 mV.

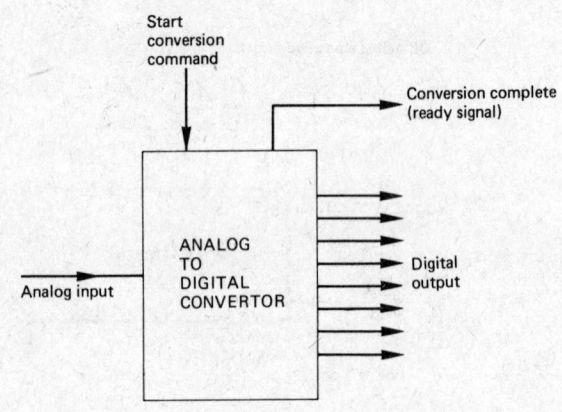
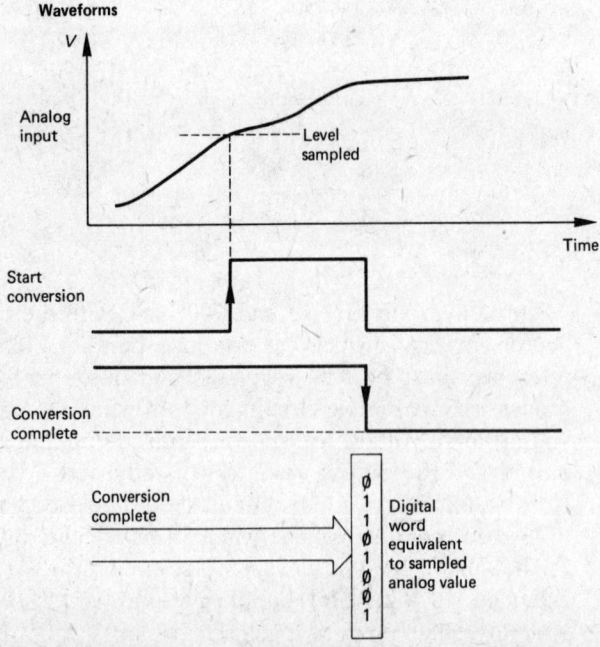

Fig. 5.16 Principle of analog-to-digital conversion

Analog input (V)	Digital output
0	000
0.35	001
0.70	010
1.05	011
1.4	100
1.75	101
2.1	110
2.45	111

From this we can see that a 1.4 V input gives the output code 100 and that when the analog input is 1.75 V the output code changes by 1 LSB to 101. But there are analog input values between 1.4 V and 1.75 V (1.575 V) which give either 100 or 101. This uncertainty is what is meant by quantising error. The larger the number of bits, the smaller the uncertainty.

So far we have assumed that the digital output of an ADC has no missing codes. A missing code for an ADC is similar to non-monotonicity in a DAC. The **useful resolution** quoted for an ADC indicates that no missing codes will be present at the digital output. The table shown above has no missing codes in the digital output and therefore the 3-bit ADC can be said to have a useful resolution of 3 bits.

Other errors that can occur are caused by **nonlinearity, offset** and **gain error**. All of these are similar to those errors specified for the DAC.

An important parameter of an ADC is the **conversion time**; the time interval between the command being given to start the conversion and the appearance at the output of the complete digital equivalent of the analog input. In many applications, speed may not be the main consideration and an ADC that is relatively slow with a conversion time of say several millisecs can be used. The speed of conversion varies with the type of ADC; this is explained in the next section.

5.5 Practical ADCs

There are many ways of performing analog-to-digital conversion. Methods range from the slow and inexpensive to the ultra fast and relatively costly. The commonly used methods are:

Voltage to frequency
Voltage to pulse width
Parallel or simultaneous conversion (flash convertors)*
Single ramp and counter*
Dual ramp
Tracking conversion*
Successive approximation*

It would probably be confusing to deal in detail with each of these and therefore only the types most applicable to micro systems will be covered. These are marked with an asterisk.

Fig. 5.17 Parallel (flash) convertor

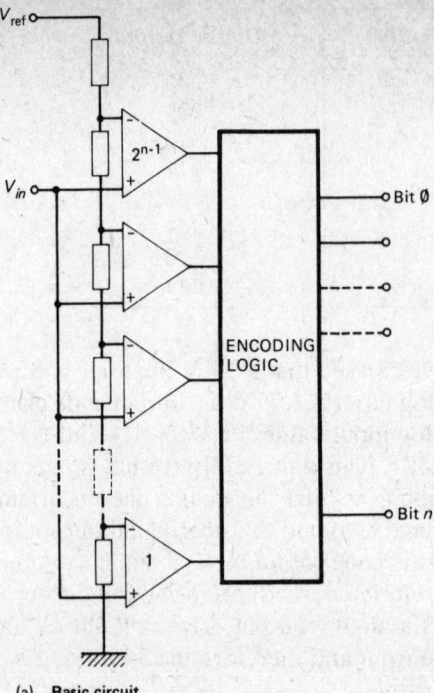

(a) Basic circuit

1 The parallel or simultaneous ADC (*fig. 5.17*)

The parallel ADC, often referred to as a *flash convertor*, is the fastest type available. This is because all the bits for the digital representation of the analog input level are determined simultaneously. The analog input is applied to a parallel bank of voltage comparators, each of which responds to a different discrete level of input voltage.

The circuit example of a 3-bit "flash" ADC gives a good idea of the complexity involved, but available types are usually 6 or 8 bits. A constant-current source supplies 1 mA to a chain of resistors R_1 to R_7. These set up the levels at which the seven comparators switch, i.e. 0.5 V, 1 V, 1.5 V, 2 V up to 3.5 V. If the analog input just exceeds 1.5 V, then the outputs of comparators A, B, and C will be logic Ø, while comparators D, E, F and G will give a high logic 1 output. By suitable logic the outputs of the comparators are then converted into the 3-bit digital output, Ø11 in this case. To get Ø11 the logic has to have the inputs

$$\overline{A}.\overline{B}.\overline{C}\,D.E.F.G$$

For *n* bits of binary output, the method requires 2^{n-1} comparators plus a lot of logic. A 6-bit convertor must have 63 comparators. The circuit complexity

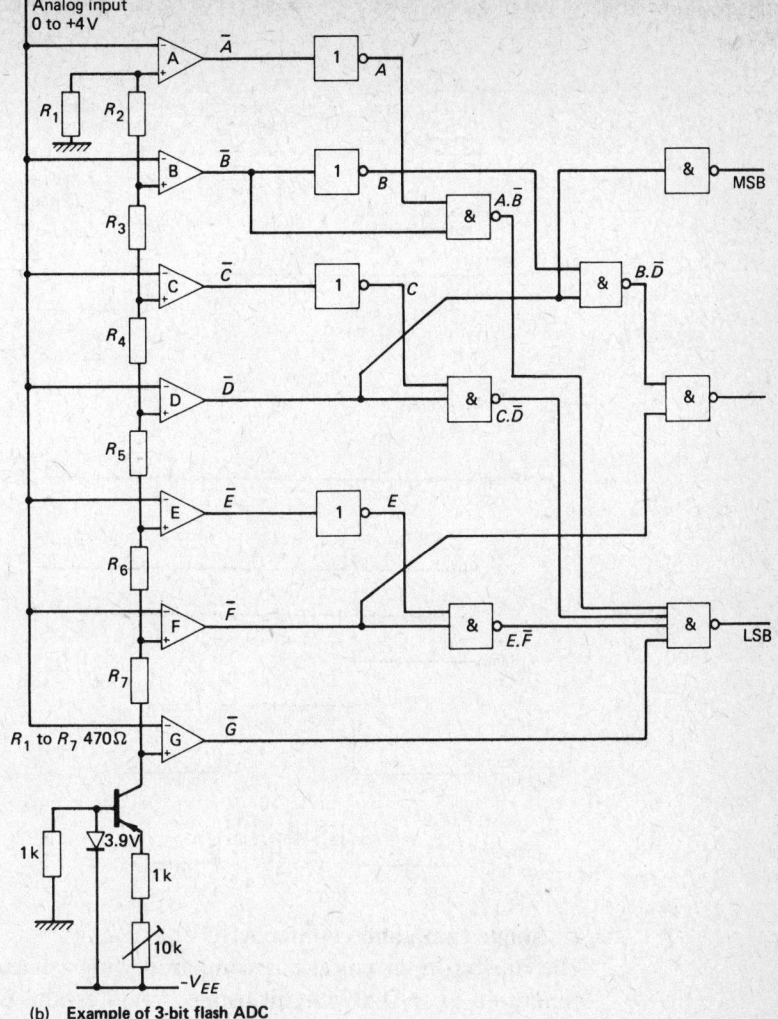

(b) **Example of 3-bit flash ADC**

increases costs which is the main disadvantage of the method, but now that LSI integrated circuits are available the method is being increasingly used, especially where speed is important. The type 3300 is an example. This 6-bit flash ADC, using high-speed CMOS logic, has a conversion speed of 15 MHz ($V_{DD} = 8\,\text{V}$), its outputs are TTL compatible, and are tri-state buffered.

Fig. 5.18 Single ramp and counter ADC

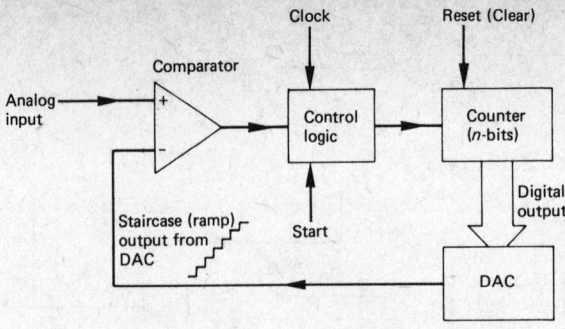

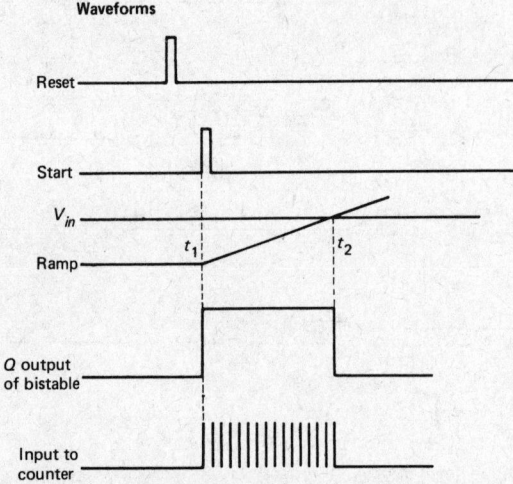

2 Single ramp and counter ADC (*fig. 5.18*)

The analog input voltage is compared with a linear ramp, the ramp being generated by a DAC circuit using a counter and R-2R network. While the input is greater than the ramp, from time t_1, the counter accumulates clock pulses. But as soon as the ramp just exceeds the input level, at time t_2, the control bistable is reset and the counter is stopped. The content of the counter is then proportional to the analog input.

This method is very useful for medium accuracy and slow speed. The conversion time, which is usually several milliseconds, is set by the clock frequency and the comparator's slew rate. Suppose a clock of 1 kHz is used, then a full 8-bit conversion will take 255 msec. The accuracy is limited by the linearity of the ramp (this can be good if the internal DAC has good linearity) and input offset drift in the comparator.

The ZN425E 8-bit dual mode A-to-D/D-to-A convertor is typical of this type. To use the IC as an ADC an external comparator and one logic IC are required. The ZN425E contains an 8-bit binary counter, an array of precision bipolar switches, an R-2R ladder network, and a 2.5 V reference. The operation is more fully described in Chapter 1 (see page 17).

Fig. 5.19 Tracking-type ADC

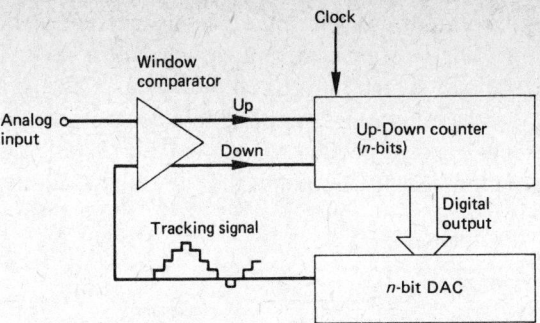

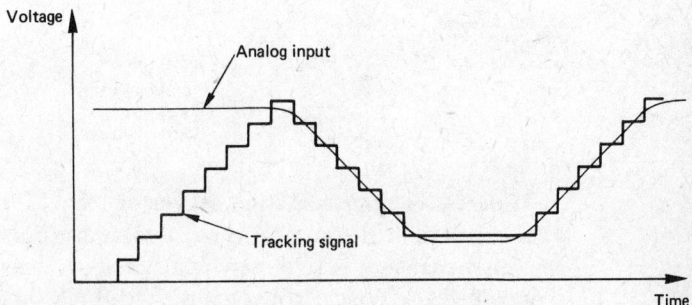

3 Tracking ADC (*fig. 5.19*)

This is a variation of the ramp and counter method that enables a changing analog input signal to be followed or "tracked". The tracking convertor uses a window comparator, one that gives changes of state if the input is outside defined levels, and an up/down counter. The counter is connected to an R-2R network so that the digital state of the counter, which is the output word, is continuously converted into analog. When the analog input is above the level from the R-2R ladder, the comparator output forces the counter to count up, causing the DAC output tracking signal to follow the analog input. If the digital value is greater than the analog input, the comparator switches in the opposite direction and causes the counter to count down. The counter is stopped when the DAC output is equal to the analog input $\pm \frac{1}{2}$LSB; the input is then within the "window" of the comparator.

A tracking convertor is inherently faster than a simple ramp and counter type because, once the first conversion, starting from zero, has been carried out, all further conversions require only that number of clock pulses necessary to track any rise or fall in the analog input. With the ramp and counter method, all conversions must be made from zero and the counter is reset between conversions. The Ferranti ZN433 ADC is one typical tracking convertor.

A similar technique is used in convertors such as the 8703 (an 8-bit CMOS ADC) where the conversion is performed by charge balancing. The analog input is applied to the IC as a current via an external resistor, and an internal amplifier integrates the sum of this current and pulses of the reference current. The reference current, which can be set up by the user, is switched by an electronic switch into the summing junction of the op-amp. The output of the op-amp integrator is applied to a comparator and the number of pulses of reference current needed to balance out the analog input current is counted. The content of the counter is then the digital equivalent of the analog input.

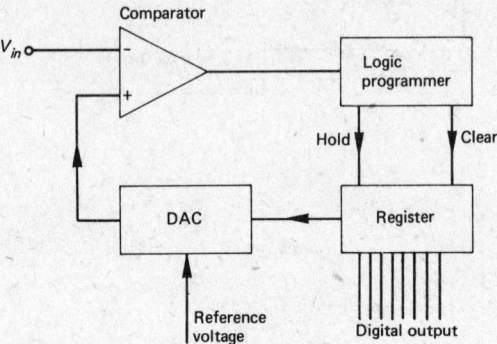

Fig. 5.20 Successive approximation ADC

4 Successive approximation convertor (fig. 5.20)

This method is the one that is most popular for use in microprocessor systems mainly because it is relatively fast, has good accuracy, and can be software-controlled. A typical conversion time for 8 bits may be only 20 μsec. The system requires a logic programmer (this part of the job can be carried out by the micro), a register to hold the result, a DAC and a comparator. At the start of the conversion, the most significant bit of the register is made logic 1 by the logic programmer so that the register (for 8 bits) reads 1ØØØØØØØ. This is converted by the DAC, and the output value compared with the analog input. If the DAC output value is larger than V_{in}, the logic 1 is removed from the most significant bit and placed in the next most significant bit (register now holds Ø1ØØØØØØ).

Suppose this gives a DAC output value that is less than V_{in}. Then the logic 1 in that position is retained and the next most significant bit is used for comparison (register content Ø11ØØØØØ). This process continues until all bits have been tried and a point of balance is reached; that is when V_{in} is just greater than the DAC output value. The method is relatively fast because the number of comparisons is equal to the number of bits used. Suppose each comparison takes 5 μsec and there are 8 bits; then the conversion time is $5 \times 8 = 40$ μsec. A successive approximation cycle, sometimes called a "put and take", is illustrated in fig. 5.21.

It will be useful at this stage to look at how a microprocessor system can be connected and programmed to give a successive approximation ADC. For this we shall consider the M6800 processor with its PIA the M6821. The mode of operations possible with this interface adaptor is covered in detail in the next chapter and therefore if you find the description too confusing to follow at present please return to this section after you have absorbed the later material.

Fig. 5.21 Successive approximation cycle

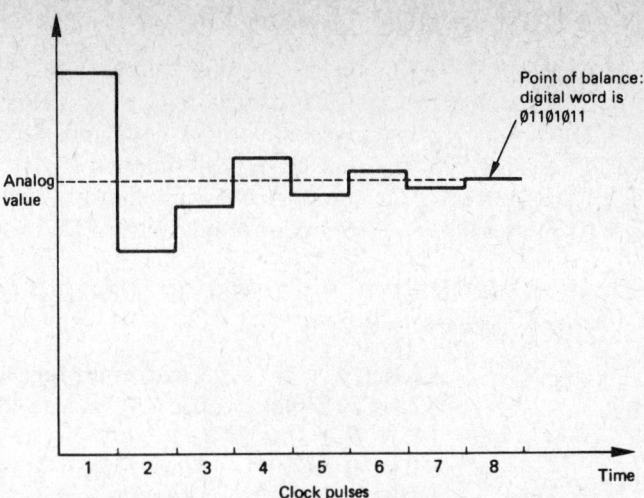

Fig. 5.22 Successive approximation using a micro

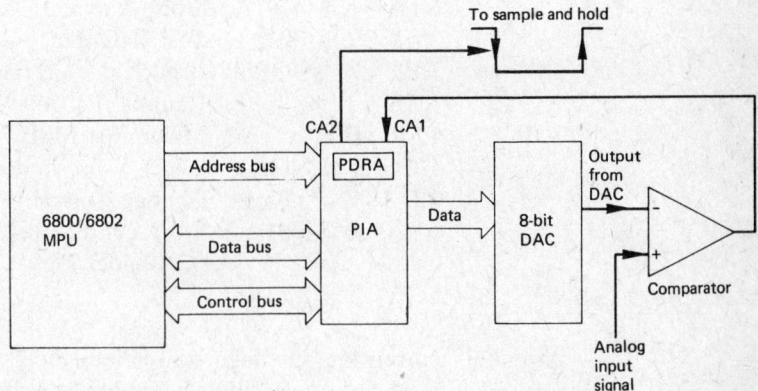

Fig. 5.23a Initialisation of PIA

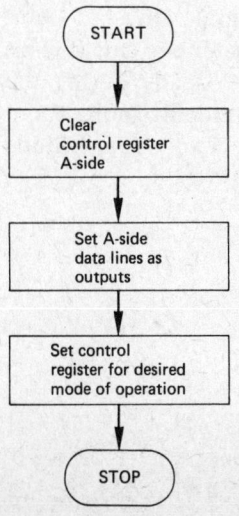

The block diagram (*fig. 5.22*) shows that only a DAC and comparator are required as extras to create the system. The M6800 micro does the job of the logic programmer and one of the data registers in the PIA becomes the SAR (successive approximation register). The operation of the circuit will be exactly as described previously. That is, the analog input is compared with the DAC output from the SAR successively, as each bit from the MSB to the LSB is tried and then retained or discarded. But for the system to operate correctly, the PIA must first be correctly initialised (set up). Here we shall use the A side of the PIA and ignore any use of the B side. The flowchart for the initialisation is shown in *fig. 5.23a* together with further explanatory comments. Note that, where instructions such as LDA A or CLR A are used, these are concerned with Accumulator A in the 6800 microprocessor and not directly with the A side registers in the PIA. The base address of the PIA is assumed to be at $4000.

INITIALISATION   *Assembly language*

CLR A	Clear Accumulator A.
STA A $4001	Clear control register A. (This gives access to DDRA.)
DEC A	Accumulator A now holds $FF.
STA A $4000	Set DDRA all outputs.
LDA A #$3C	Load Acc A with control word.
STA A $4001	Set control register in PIA.

SUBROUTINE FOR SUCCESSIVE APPROXIMATION
(see Flowchart *fig. 5.23b*)

START	CLR	A	Clear result register.
	STA	A $4000	Clear PDRA (SAR).
	LDA	B #$34	
	STA	B $4001	Store control word (CA2 goes low).
	LDA	B #$80	Set MSB = 1.
LOOP 1	ABA		Build result in Acc A.
	STA	A $4000	Store A in PDRA (SAR).
	TST	$4001	Test IRQ flag.
	BPL	LOOP 2	Branch if $V_{IN} > V_{DAC}$.
	SBA		Remove 1 from Acc A.
LOOP 2	LSR	B	Make next MSB = 1.
	BCC	LOOP 1	Check if finished.
	LDA	B #$3C	Change control word.
	STA	B $4001	Store control word (CA2 goes high).
END	RTS		

After this subroutine has been completed, the digital word equivalent to the value of the analog input will be held both in Accumulator A in the M6800 and in the A side peripheral data register in the PIA. This word could be used by the main program or stored in memory until required. One last point concerns the use of CA2 as a timing control signal for a sampling circuit. This may not be required if the analog input is relatively steady during the conversion time (about 100 μsec). A sample-and-hold circuit (more about that in the next section) is essential if the analog input does vary in amplitude during the time it takes for the conversion to be carried out.

Conversion 125

Fig. 5.23b Successive approximation routine

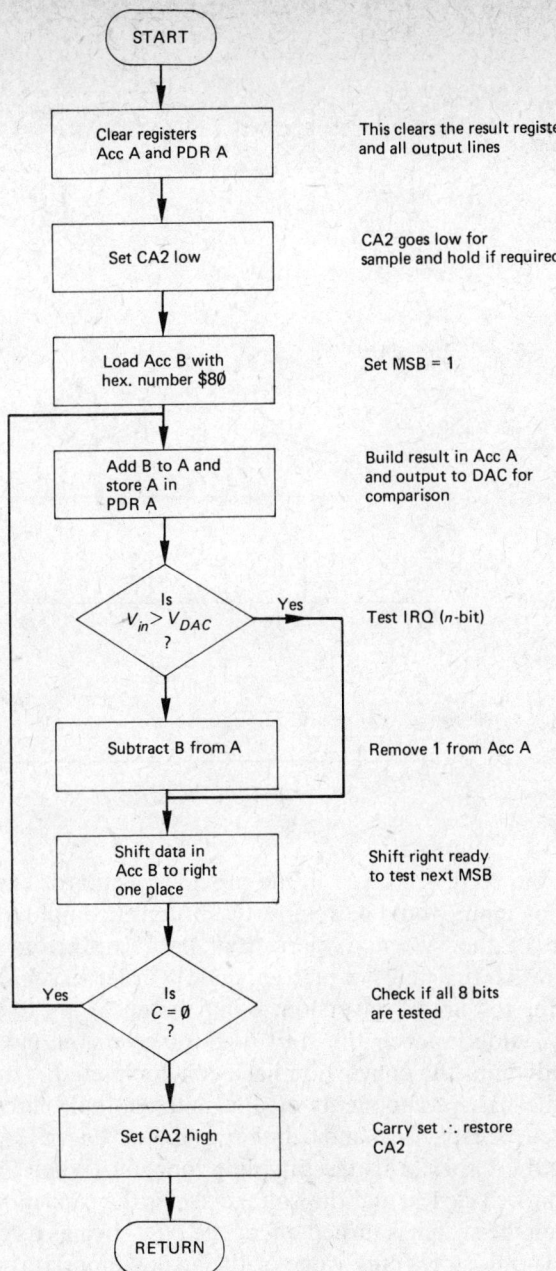

Fig. 5.24 Principle of sample-and-hold circuit

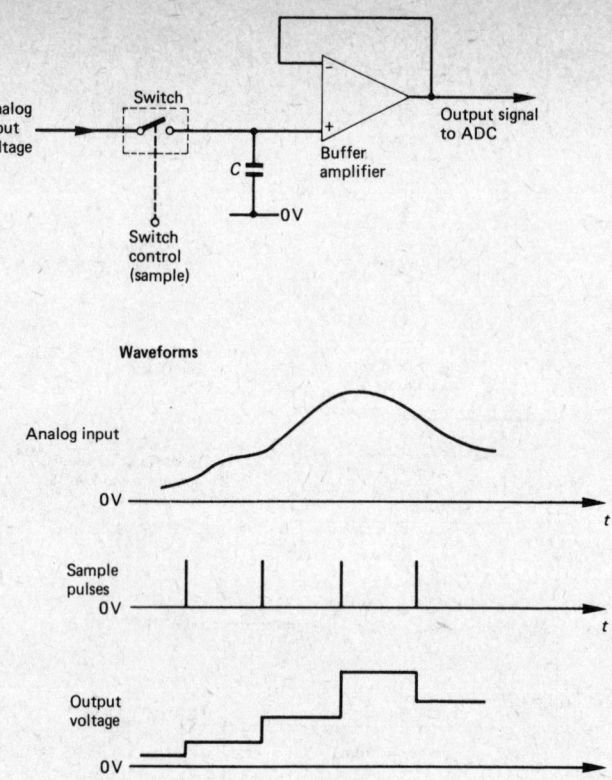

5.6 Sample-and-Hold Circuits

For the ADCs covered in the previous section it has been assumed that the analog input would be relatively constant in amplitude during the conversion time of the system. Apart from flash convertors which are very fast in operation this may not be true, and, in order to achieve reasonably error-free analog-to-digital conversion, it may be necessary to sample the amplitude of the analog input at the start of the conversion and then to hold this value steady until the conversion has been completed.

The basic components of a sample-and-hold circuit are a fast electronic switch, a capacitor, and a buffer amplifier (*fig. 5.24*). The switch, usually a MOSFET, performs the sampling function. When the control signal is high, the switch closes, and the voltage across the capacitor tracks the input signal. When the switch is turned off, as the ADC begins its conversion, the capacitor holds (i.e. stores) the value of the analog input at that instant. At the end of the conversion the switch is again closed and the capacitor charges to any new value of analog input. The buffer amplifier has a very high input resistance and therefore prevents the capacitor from being discharged during the time that the switch is open. To ensure that the sampled analog input is held effectively, the capacitor must be a high-quality low-leakage type; usually a polypropylene or polystyrene are best. There are several sample-and-hold ICs available which only require an external capacitor to be fitted.

6 Microprocessors and Interface Adaptors

6.1 Types of Interface Adaptor

An IC which is designed to do the task of matching a microprocessor to peripheral equipment is called an **interface adaptor** (*fig. 6.1*). Many ICs have therefore been developed by manufacturers specifically for certain processors. For example: the 6502 micro has the interface adaptors 6522 VIA (versatile interface adaptor), 6532 RIOT (RAM, I/O, interval timer), and the 6551 ACIA (asynchronous communications interface adaptor). The Z80 has the Z80 PIO (peripheral input/output controller) and the Z80A SIO (serial input/output controller). The 6800 has, amongst others, the 6821 PIA (peripheral interface adaptor) and the 6850 ACIA (asynchronous communications interface adaptor).

Fig. 6.1 General view of an interface adaptor

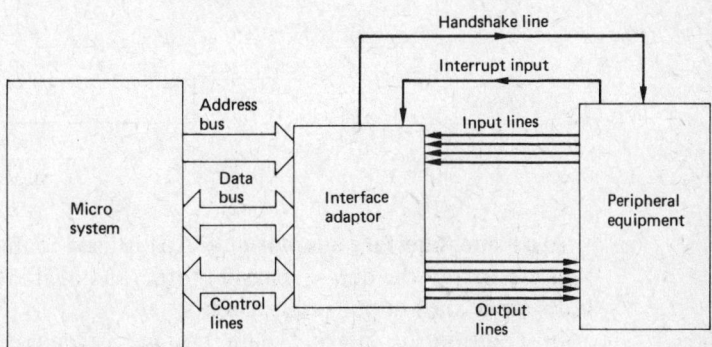

These devices are usually labelled as input/output (I/O) on a system block diagram and are memory mapped. In other words, access to registers inside the adaptor can be made by calling up a particular address. Although each IC usually has many other extra facilities, and are given different names by manufacturers, these interface adaptors are mainly of two types; that is, either serial or parallel.

Fig. 6.2 Serial interface adaptor

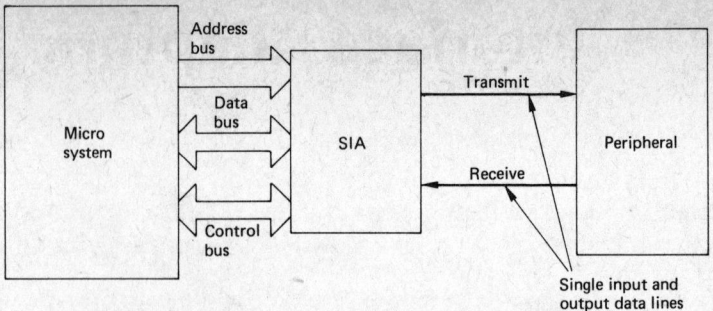

A **serial interface adaptor** (*fig. 6.2*) is designed to interface with peripherals that accept data one bit at a time as a train of pulses along a single wire. The SIA therefore has the job of converting the parallel data from the microprocessor into serial and to output this data at a rate that can be accepted by the peripheral (a teletype for example). It also may be required to take serial data as an input and convert this into parallel data for the micro.

Fig. 6.3 Parallel interface adaptor

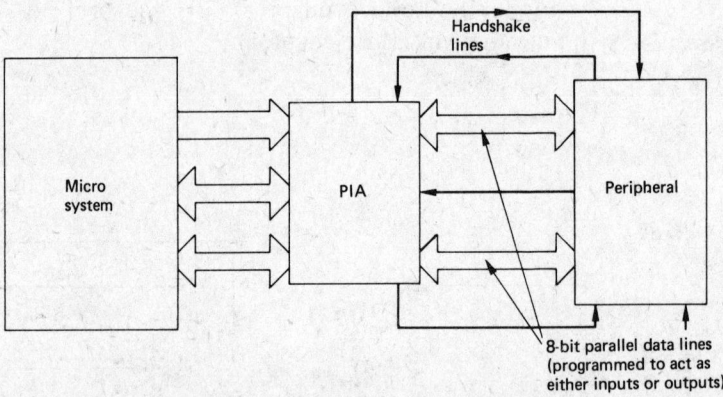

A **parallel interface adaptor** (*fig. 6.3*) is used to match the micro to devices that can accept the data in parallel form, and again the adaptor will deal with both input and output requirements.

In addition to internal data storage registers, the adaptor will have handshaking facilities and control registers. It will not be possible to describe in detail all the major interface adaptors available for all the different processors, a choice has to be made, and this chapter deals with the 6800 series processor and its interface adaptors.

6.2 The 6800 Range of Microprocessors

We shall be examining mainly the basic 6800 processor, the 6821 PIA, and the 6850 ACIA, but before doing that it will be useful to look at the family of ICs available in the 6800 range (*fig. 6.4*).

The 8-bit microprocessor **MC6800** forms the starting point for the rest of the 6800 family. All later processors in this range are based on the

Fig. 6.4 The 6800 family of microprocessors and microcomputers

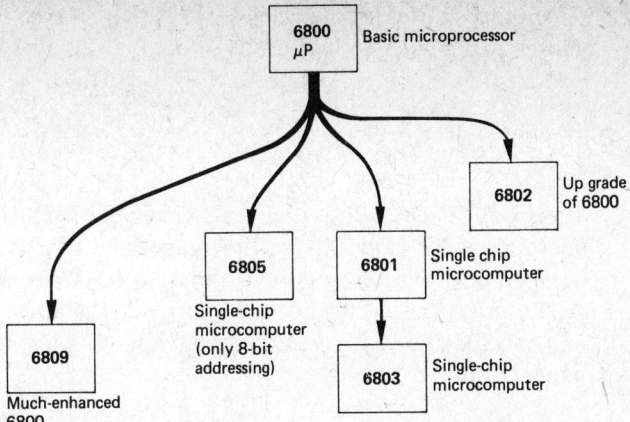

architecture and instruction set of the MC6800. It is a highly cost effective device ideally suited to process control applications. Briefly the main points about it are

TTL compatible and requires only one +5 V power supply.
16-bit address bus—64 K bytes of addressing.
72 instructions.
7 addressing modes.
6 internal registers—two accumulators, index register, program counter, stack pointer, and condition code register (status register).

An external clock generator (6875) is required to provide the ϕ_1 and ϕ_2 clock signals and the most basic unit operates at 1 MHz; but 1.5 MHz and 2.0 MHz versions (MC68A00 and MC68B00) are available.

The more powerful **6801** is a single-chip microcomputer; in other words it has a 6800 processor together with a clock generator, 128 bytes of RAM, 2 K bytes of ROM, a serial communications interface (SC1), several parallel I/O lines, and a three function programmable timer, all inside one 40-pin package. In addition, the 6801 possesses several new instructions and some of the key instructions have faster execution times than the 6800.

The **6802** is an up-graded 6800. It has all the registers and features of the 6800 plus an internal clock circuit and 128 bytes of RAM (located at addresses $0000 to $007F). The 6802 is therefore recommended for new designs.

Other versions include the 6803 and 6805. The **6803** is similar to the 6801 but without the on-board ROM, whereas the **6805**, housed in a 28-pin package, has shortened registers and reduced memory addressing capability (it has only 8 address lines instead of 16). But with its 64 bytes of RAM, 1 K bytes of user ROM, the TTL/CMOS compatible I/O lines, and on-chip clock it does provide a very economical microcomputer system. A CMOS version (MC146805) and an EPROM version (MC 68705) are also available.

The **6809** 8-bit microprocessor is a much-enhanced 6800 with major architectural improvements, more instructions, and more addressing modes. For example, it has two 16-bit index registers and two 16-bit indexable stack pointers.

Another feature of the 6809 is that its Direct Page Register allows the direct addressing mode to be used throughout memory rather than for only addresses $0000 to $00FF as in the basic 6800.

6.3 Architecture and Programming the 6800/6802

There are of course several texts that deal exclusively with the 6800/6802 microprocessor and this section can only give brief coverage. For more detailed information see the *Motorola 8-bit Microprocessor Data Manual*.

The internal structure of the processor (6800) showing the main registers and control signals is given in *fig. 6.5* and a typical minimum system set-up is indicated in *fig. 6.6*.

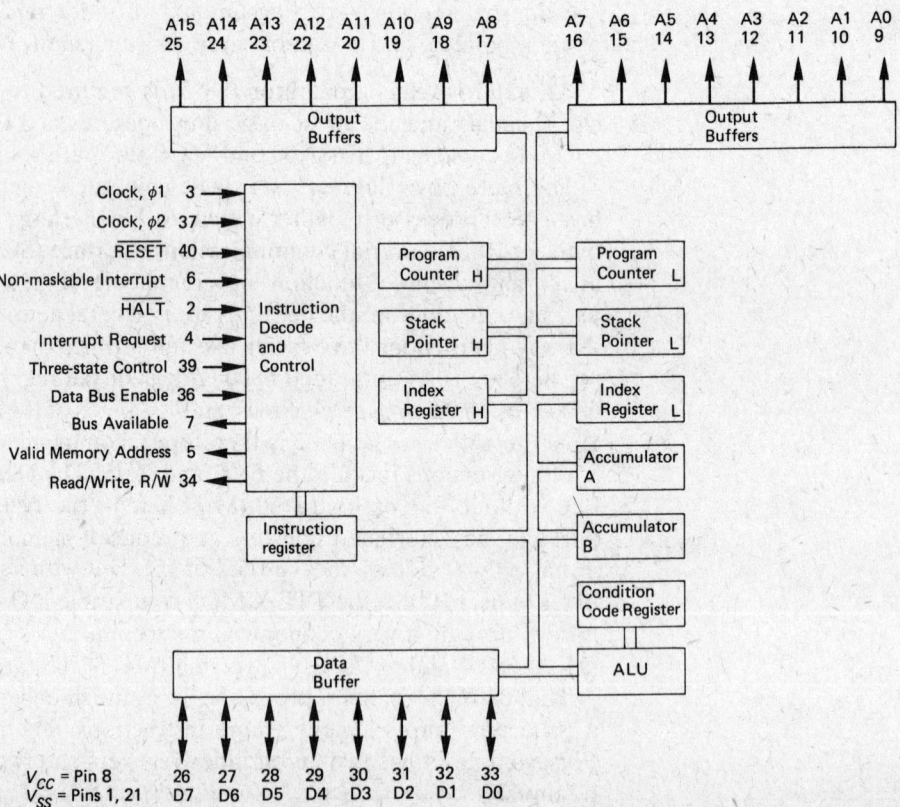

Fig. 6.5a Architecture of the 6800

Fig. 6.5b Pin-out of the 6800

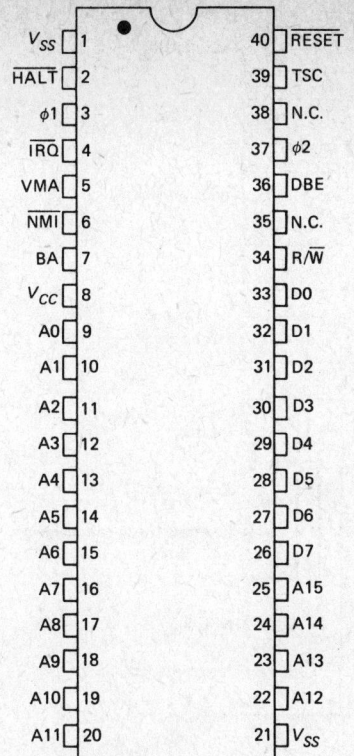

Fig. 6.6 Typical minimum system set-up; such a system can easily be adapted for a number of small-scale applications by changing the application program content of the ROM

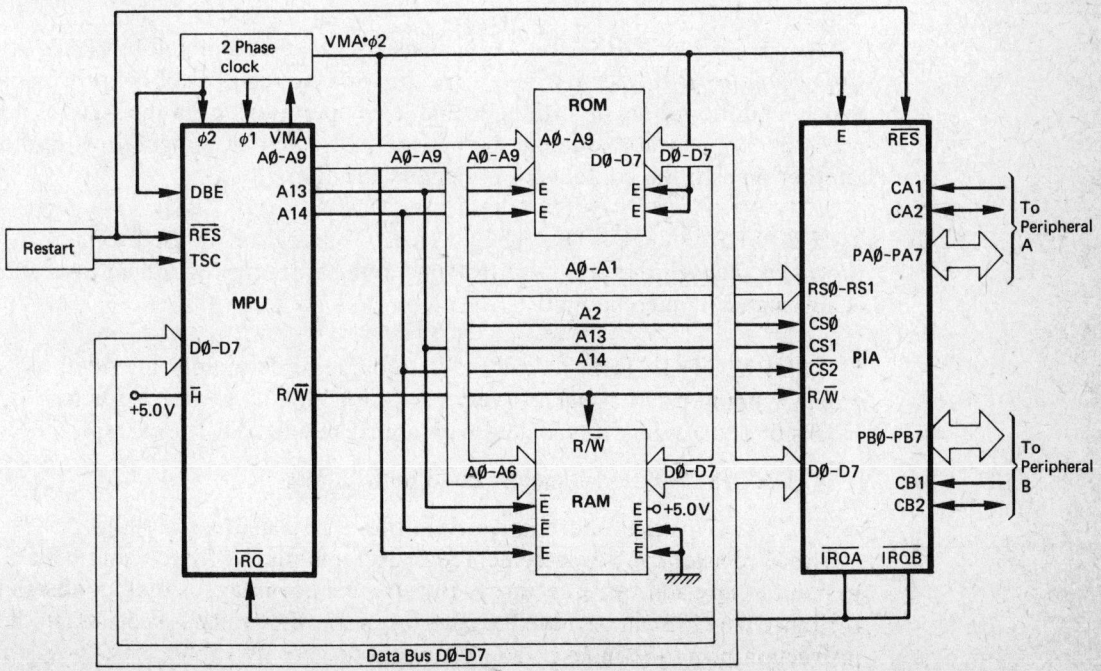

Fig. 6.7 Programming model of the 6800 microprocessor

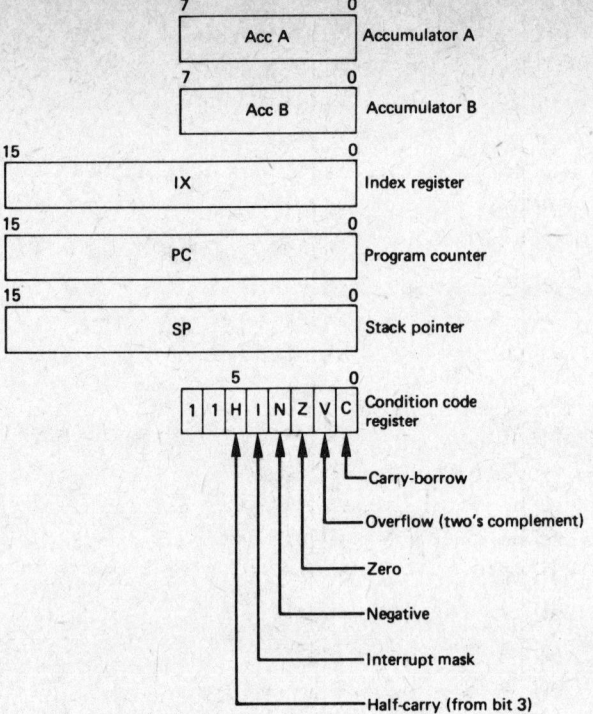

The register set is as follows (*fig. 6.7*):

1 ACCUMULATORS Two 8-bit registers called accumulator A (Acc A) and accumulator B (Acc B). These are the working registers of the processor which hold the results of arithmetic and logic operations from the ALU. The accumulators can, for example, be added, subtracted or compared with one another or with the contents of a memory location.

2 PROGRAM COUNTER (PC) This 16-bit register is used to step the processor through the program. It points to the current program address and is automatically incremented.

3 INDEX REGISTER (X reg) A 16-bit register mainly intended as a memory pointer and as such is very useful for look-up tables or for moving blocks of data. When it is loaded with a hex, number, as for example,

  LDX #$1∅∅∅   (# means Load immediate)

it will "point" to that address, which in this case is address $1∅∅∅.

The X register can be decremented or incremented so that it can be made to point to any address in memory. But the real advantage is that an **offset** in the range 0 to 255 can be added to its value. The offset value is specified in the instruction using the indexed addressing mode. For example:

Program instruction	*Comments*
LDX #$15ØØ	X register set to point to address $15ØØ.
LDA A ØØ,X	Acc A loaded from "pointed" address, i.e. from address $15ØØ (offset is zero).
STA A Ø5,X	Contents of Acc A stored at an address which equals pointed address plus offset of Ø5, i.e. address $15Ø5.

The use of this offset gives flexibility for the indexed addressing mode.

4 STACK POINTER (SP) Another 16-bit register which contains the address of the next available location in an area of memory called the stack. The **stack** is a last-in/first-out portion of RAM defined by the initial hex. value loaded into the stack pointer. The stack concept is important for subroutines and interrupts.

When a JSR (jump to subroutine) or BSR (branch to subroutine) instruction appears in the program, the return address is automatically saved on the stack. This is retrieved when an RTS (return from subroutine) instruction is executed.

When an interrupt occurs from some peripheral device, processor status is automatically saved on the stack. Processor status for the 6800 means the content of all the internal registers, i.e. the condition code register, Acc B, Acc A, X register, and PC. After the interrupt service routine has been executed, status is restored to the processor so that it can continue with the main program (more on interrupts later).

5 CONDITION CODE REGISTER (CCR) This register contains a group of five flags which are set or reset according to the result of logic or arithmetic operations. The five flags are shown in *fig. 6.7* and are used mainly in conditional branching when the state of one, or a combination, of the flags is checked by the processor, and a program branch is executed if certain conditions are met. For example:

BCS	Branch if carry set ($C=1$)
BCC	Branch if carry clear ($C=0$)
BEQ	Branch if equal (to zero) ($Z=1$)
BNE	Branch if not equal (to zero) ($Z=0$)

Instruction Set and Addressing

The 6800/6802 has a set of 72 instructions but since there are seven addressing modes the total number of different instructions adds up to 197. The table in *fig. 6.8* gives the instruction set, and it can be seen that instructions can be one, two or three bytes in length. The first or only byte identifies the instruction and the other bytes contain either data or an address. For example,

Single byte   SBA   Subtract Accumulators
Double byte   LDA B $0050   Load Acc B from address $50
Triple byte   LDS #$3500   Load stack pointer with hex. number $3500

The various **addressing modes**, which refer to the way in which the program causes the processor to obtain its instructions or data, are:

Immediate   IMM
Direct   DIR
Extended   EXT
Indexed   IND
Relative   REL
Inherent   (sometimes called Implied)
Accumulator   A and B

Immediate   In the Immediate addressing mode, the data immediately following the instruction is the value which is to be operated on (i.e. the *operand*). Take the example

Operator   *Operand*   *Comments*
LDA B   #$FF   Load $FF into Acc B

Here the instructions causes the hex. number FF to be loaded immediately into Acc B.

Direct   With this addressing mode, a single-byte page zero address is specified as part of the instruction. Example:

Operator   *Operand*   *Comments*
LDA $0080   Load Acc A from address $80

This addressing mode can only be used for address locations from $0000 to $00FF.

Extended   This enables any address in memory to be used. A two-byte address is specified. Examples:

   STX $3150   Store index register at address $3150
   LDA A $20F0   Load Acc A from address $20F0

Indexed   This uses the index register as a pointer with an offset supplied with the instruction. The offset is positive in the range $00 to $FF (0 to 255).
     Examples (assume X reg. holds hex. value $2000):

   STA A $00,X   Store Acc A at address $2000 (offset is 00)
   STA B $FF,X   Store Acc B at address $20FF (offset is $FF)

Relative This addressing mode is used in branch instructions and allows branching both forwards and backwards in memory relative to the value held by the program counter. The range is from −125 to +129 locations relative to the first byte of the two-byte relative branch instruction.

Inherent (sometimes called Implied) Instructions such as DEX (decrement index register) and TSX (transfer from stack pointer to index register) are those which operate directly on registers within the processor and no address or data has to be supplied. This is called the inherent addressing mode.

Accumulator An extension of inherent addressing where the instructions are those which cause changes to the accumulators and therefore A or B must be specified. For example,

    INC A   Increment Acc A
    INC B   Increment Acc B
    CLR A   Clear Acc A
    CLR B   Clear Acc B

To complete the picture we need to look at **control signals** and in particular to see how interrupts are handled. Refer back to *fig. 6.5*.

Clock signals, phase 1 (ϕ_1) and phase 2 (ϕ_2) Two pins (3,37) are used for a two-phase non-overlapping clock. These signals synchronise the operation of the processor.

Reset $\overline{\text{RES}}$ This input is used to reset and start the processor, either as initial start-up or following a power-down condition.

If a high level is detected on this input, the program counter is loaded with the address stored in the restart pointer ($FFFE and $FFFF) and the processor then proceeds with the execution of a Restart Program. This restart depends, however, on the $\overline{\text{HALT}}$ control input being in a high state.

Non-maskable interrupt $\overline{\text{NMI}}$ This input, mostly used for such events as power failure or a fault condition, causes an immediate interrupt to the processor. If a negative edge is detected on this control input, the current instruction will be executed, processor status will be saved on the stack, and the program counter will be loaded from the non-maskable interrupt pointer (address $FFFC and $FFFD).

$\overline{\text{HALT}}$ When this level sensitive input is in the low state, all activity in the machine will stop. This allows control of program execution by an outside source. Normally this pin is connected to +5 V.

Fig. 6.8 The 6800 Instruction Set

ACCUMULATOR AND MEMORY INSTRUCTIONS

OPERATIONS	MNEMONIC	IMMED OP ~ =	DIRECT OP ~ =	INDEX OP ~ =	EXTND OP ~ =	IMPLIED OP ~ =	BOOLEAN/ARITHMETIC OPERATION (All register labels refer to contents)	COND. CODE REG. 5 4 3 2 1 0 H I N Z V C
Add	ADDA	8B 2 2	9B 3 2	AB 5 2	BB 4 3		A + M → A	↕ • ↕ ↕ ↕ ↕
	ADDB	CB 2 2	DB 3 2	EB 5 2	FB 4 3		B + M → B	↕ • ↕ ↕ ↕ ↕
Add Acmltrs	ABA					1B 2 1	A + B → A	↕ • ↕ ↕ ↕ ↕
Add with Carry	ADCA	89 2 2	99 3 2	A9 5 2	B9 4 3		A + M + C → A	↕ • ↕ ↕ ↕ ↕
	ADCB	C9 2 2	D9 3 2	E9 5 2	F9 4 3		B + M + C → B	↕ • ↕ ↕ ↕ ↕
And	ANDA	84 2 2	94 3 2	A4 5 2	B4 4 3		A · M → A	• • ↕ ↕ R •
	ANDB	C4 2 2	D4 3 2	E4 5 2	F4 4 3		B · M → B	• • ↕ ↕ R •
Bit Test	BITA	85 2 2	95 3 2	A5 5 2	B5 4 3		A · M	• • ↕ ↕ R •
	BITB	C5 2 2	D5 3 2	E5 5 2	F5 4 3		B · M	• • ↕ ↕ R •
Clear	CLR			6F 7 2	7F 6 3		00 → M	• • R S R R
	CLRA					4F 2 1	00 → A	• • R S R R
	CLRB					5F 2 1	00 → B	• • R S R R
Compare	CMPA	81 2 2	91 3 2	A1 5 2	B1 4 3		A − M	• • ↕ ↕ ↕ ↕
	CMPB	C1 2 2	D1 3 2	E1 5 2	F1 4 3		B − M	• • ↕ ↕ ↕ ↕
Compare Acmltrs	CBA					11 2 1	A − B	• • ↕ ↕ ↕ ↕
Complement, 1's	COM			63 7 2	73 6 3		$\overline{M}$ → M	• • ↕ ↕ R S
	COMA					43 2 1	$\overline{A}$ → A	• • ↕ ↕ R S
	COMB					53 2 1	$\overline{B}$ → B	• • ↕ ↕ R S
Complement, 2's (Negate)	NEG			60 7 2	70 6 3		00 − M → M	• • ↕ ↕ ① ②
	NEGA					40 2 1	00 − A → A	• • ↕ ↕ ① ②
	NEGB					50 2 1	00 − B → B	• • ↕ ↕ ① ②
Decimal Adjust, A	DAA					19 2 1	Converts Binary Add. of BCD Characters into BCD Format	• • ↕ ↕ ↕ ③
Decrement	DEC			6A 7 2	7A 6 3		M − 1 → M	• • ↕ ↕ 4 •
	DECA					4A 2 1	A − 1 → A	• • ↕ ↕ 4 •
	DECB					5A 2 1	B − 1 → B	• • ↕ ↕ 4 •
Exclusive OR	EORA	88 2 2	98 3 2	A8 5 2	B8 4 3		A ⊕ M → A	• • ↕ ↕ R •
	EORB	C8 2 2	D8 3 2	E8 5 2	F8 4 3		B ⊕ M → B	• • ↕ ↕ R •
Increment	INC			6C 7 2	7C 6 3		M + 1 → M	• • ↕ ↕ ⑤ •
	INCA					4C 2 1	A + 1 → A	• • ↕ ↕ ⑤ •
	INCB					5C 2 1	B + 1 → B	• • ↕ ↕ ⑤ •
Load Acmltr	LDAA	86 2 2	96 3 2	A6 5 2	B6 4 3		M → A	• • ↕ ↕ R •
	LDAB	C6 2 2	D6 3 2	E6 5 2	F6 4 3		M → B	• • ↕ ↕ R •
Or, Inclusive	ORAA	8A 2 2	9A 3 2	AA 5 2	BA 4 3		A + M → A	• • ↕ ↕ R •
	ORAB	CA 2 2	DA 3 2	EA 5 2	FA 4 3		B + M → B	• • ↕ ↕ R •
Push Data	PSHA					36 4 1	A → M$_{SP}$, SP − 1 → SP	• • • • • •
	PSHB					37 4 1	B → M$_{SP}$, SP − 1 → SP	• • • • • •
Pull Data	PULA					32 4 1	SP + 1 → SP, M$_{SP}$ → A	• • • • • •
	PULB					33 4 1	SP + 1 → SP, M$_{SP}$ → B	• • • • • •
Rotate Left	ROL			69 7 2	79 6 3		M ⎤	• • ↕ ↕ ⑥ ↕
	ROLA					49 2 1	A 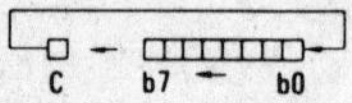 ⎬ C ← b7 ← b0	• • ↕ ↕ ⑥ ↕
	ROLB					59 2 1	B ⎦	• • ↕ ↕ ⑥ ↕
Rotate Right	ROR			66 7 2	76 6 3		M ⎤	• • ↕ ↕ ⑥ ↕
	RORA					46 2 1	A 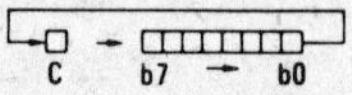 ⎬ C → b7 → b0	• • ↕ ↕ ⑥ ↕
	RORB					56 2 1	B ⎦	• • ↕ ↕ ⑥ ↕
Shift Left, Arithmetic	ASL			68 7 2	78 6 3		M ⎤	• • ↕ ↕ ⑥ ↕
	ASLA					48 2 1	A 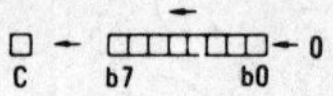 ⎬ ← b7 ← b0 ← 0	• • ↕ ↕ ⑥ ↕
	ASLB					58 2 1	B ⎦ C	• • ↕ ↕ ⑥ ↕
Shift Right, Arithmetic	ASR			67 7 2	77 6 3		M ⎤	• • ↕ ↕ ⑥ ↕
	ASRA					47 2 1	A  ⎬ → → C	• • ↕ ↕ ⑥ ↕
	ASRB					57 2 1	B ⎦ b7 b0 C	• • ↕ ↕ ⑥ ↕
Shift Right, Logic	LSR			64 7 2	74 6 3		M ⎤	• • R ↕ ⑥ ↕
	LSRA					44 2 1	A  ⎬ 0 → →	• • R ↕ ⑥ ↕
	LSRB					54 2 1	B ⎦ b7 b0 C	• • R ↕ ⑥ ↕
Store Acmltr.	STAA		97 4 2	A7 6 2	B7 5 3		A → M	• • ↕ ↕ R •
	STAB		D7 4 2	E7 6 2	F7 5 3		B → M	• • ↕ ↕ R •
Subtract	SUBA	80 2 2	90 3 2	A0 5 2	B0 4 3		A − M → A	• • ↕ ↕ ↕ ↕
	SUBB	C0 2 2	D0 3 2	E0 5 2	F0 4 3		B − M → B	• • ↕ ↕ ↕ ↕
Subtract Acmltrs.	SBA					10 2 1	A − B → A	• • ↕ ↕ ↕ ↕
Subtr. with Carry	SBCA	82 2 2	92 3 2	A2 5 2	B2 4 3		A − M − C → A	• • ↕ ↕ ↕ ↕
	SBCB	C2 2 2	D2 3 2	E2 5 2	F2 4 3		B − M − C → B	• • ↕ ↕ ↕ ↕
Transfer Acmltrs	TAB					16 2 1	A → B	• • ↕ ↕ R •
	TBA					17 2 1	B → A	• • ↕ ↕ R •
Test, Zero or Minus	TST			6D 7 2	7D 6 3		M − 00	• • ↕ ↕ R R
	TSTA					4D 2 1	A − 00	• • ↕ ↕ R R
	TSTB					5D 2 1	B − 00	• • ↕ ↕ R R
								H I N Z V C

INDEX REGISTER AND STACK MANIPULATION INSTRUCTIONS

POINTER OPERATIONS	MNEMONIC	IMMED OP	~	#	DIRECT OP	~	#	INDEX OP	~	#	EXTND OP	~	#	IMPLIED OP	~	#	BOOLEAN/ARITHMETIC OPERATION	COND. CODE REG. 5 H	4 I	3 N	2 Z	1 V	0 C
Compare Index Reg	CPX	8C	3	3	9C	4	2	AC	6	2	BC	5	3				$X_H - M, X_L - (M+1)$	•	•	⑦	↕	⑧	•
Decrement Index Reg	DEX													09	4	1	$X - 1 \rightarrow X$	•	•	•	↕	•	•
Decrement Stack Pntr	DES													34	4	1	$SP - 1 \rightarrow SP$	•	•	•	•	•	•
Increment Index Reg	INX													08	4	1	$X + 1 \rightarrow X$	•	•	•	↕	•	•
Increment Stack Pntr	INS													31	4	1	$SP + 1 \rightarrow SP$	•	•	•	•	•	•
Load Index Reg	LDX	CE	3	3	DE	4	2	EE	6	2	FE	5	3				$M \rightarrow X_H, (M+1) \rightarrow X_L$	•	•	⑨	↕	R	•
Load Stack Pntr	LDS	8E	3	3	9E	4	2	AE	6	2	BE	5	3				$M \rightarrow SP_H, (M+1) \rightarrow SP_L$	•	•	⑨	↕	R	•
Store Index Reg	STX				DF	5	2	EF	7	2	FF	6	3				$X_H \rightarrow M, X_L \rightarrow (M+1)$	•	•	⑨	↕	R	•
Store Stack Pntr	STS				9F	5	2	AF	7	2	BF	6	3				$SP_H \rightarrow M, SP_L \rightarrow (M+1)$	•	•	⑨	↕	R	•
Indx Reg → Stack Pntr	TXS													35	4	1	$X - 1 \rightarrow SP$	•	•	•	•	•	•
Stack Pntr → Indx Reg	TSX													30	4	1	$SP + 1 \rightarrow X$	•	•	•	•	•	•

JUMP AND BRANCH INSTRUCTIONS

OPERATIONS	MNEMONIC	RELATIVE OP	~	#	INDEX OP	~	#	EXTND OP	~	#	IMPLIED OP	~	#	BRANCH TEST	5 H	4 I	3 N	2 Z	1 V	0 C
Branch Always	BRA	20	4	2										None	•	•	•	•	•	•
Branch If Carry Clear	BCC	24	4	2										C = 0	•	•	•	•	•	•
Branch If Carry Set	BCS	25	4	2										C = 1	•	•	•	•	•	•
Branch If = Zero	BEQ	27	4	2										Z = 1	•	•	•	•	•	•
Branch If ≥ Zero	BGE	2C	4	2										$N \oplus V = 0$	•	•	•	•	•	•
Branch If > Zero	BGT	2E	4	2										$Z + (N \oplus V) = 0$	•	•	•	•	•	•
Branch If Higher	BHI	22	4	2										C + Z = 0	•	•	•	•	•	•
Branch If ≤ Zero	BLE	2F	4	2										$Z + (N \oplus V) = 1$	•	•	•	•	•	•
Branch If Lower Or Same	BLS	23	4	2										C + Z = 1	•	•	•	•	•	•
Branch If < Zero	BLT	2D	4	2										$N \oplus V = 1$	•	•	•	•	•	•
Branch If Minus	BMI	2B	4	2										N = 1	•	•	•	•	•	•
Branch If Not Equal Zero	BNE	26	4	2										Z = 0	•	•	•	•	•	•
Branch If Overflow Clear	BVC	28	4	2										V = 0	•	•	•	•	•	•
Branch If Overflow Set	BVS	29	4	2										V = 1	•	•	•	•	•	•
Branch If Plus	BPL	2A	4	2										N = 0	•	•	•	•	•	•
Branch To Subroutine	BSR	8D	8	2											•	•	•	•	•	•
Jump	JMP				6E	4	2	7E	3	3				} See Special Operations	•	•	•	•	•	•
Jump To Subroutine	JSR				AD	8	2	BD	9	3					•	•	•	•	•	•
No Operation	NOP										01	2	1	Advances Prog. Cntr. Only	•	•	•	•	•	•
Return From Interrupt	RTI										3B	10	1		⑩
Return From Subroutine	RTS										39	5	1	} See Special Operations	•	•	•	•	•	•
Software Interrupt	SWI										3F	12	1		•	S	•	•	•	•
Wait for Interrupt*	WAI										3E	9	1		•	⑪	•	•	•	•

*WAI puts Address Bus, R/W, and Data Bus in the three-state mode while VMA is held low.

CONDITION CODE REGISTER MANIPULATION INSTRUCTIONS

OPERATIONS	MNEMONIC	IMPLIED OP	~	#	BOOLEAN OPERATION	COND. CODE REG. 5 H	4 I	3 N	2 Z	1 V	0 C
Clear Carry	CLC	0C	2	1	$0 \rightarrow C$	•	•	•	•	•	R
Clear Interrupt Mask	CLI	0E	2	1	$0 \rightarrow I$	•	R	•	•	•	•
Clear Overflow	CLV	0A	2	1	$0 \rightarrow V$	•	•	•	•	R	•
Set Carry	SEC	0D	2	1	$1 \rightarrow C$	•	•	•	•	•	S
Set Interrupt Mask	SEI	0F	2	1	$1 \rightarrow I$	•	S	•	•	•	•
Set Overflow	SEV	0B	2	1	$1 \rightarrow V$	•	•	•	•	S	•
Acmltr A → CCR	TAP	06	2	1	$A \rightarrow CCR$	⑫
CCR → Acmltr A	TPA	07	2	1	$CCR \rightarrow A$	•	•	•	•	•	•

CONDITION CODE REGISTER NOTES: (Bit set if test is true and cleared otherwise)

1	(Bit V)	Test: Result = 10000000?
2	(Bit C)	Test: Result ≠ 00000000?
3	(Bit C)	Test: Decimal value of most significant BCD Character greater than nine? (Not cleared if previously set.)
4	(Bit V)	Test: Operand = 10000000 prior to execution?
5	(Bit V)	Test: Operand = 01111111 prior to execution?
6	(Bit V)	Test: Set equal to result of N⊕C after shift has occurred.
7	(Bit N)	Test: Sign bit of most significant (MS) byte = 1?
8	(Bit V)	Test: 2's complement overflow from subtraction of MS bytes?
9	(Bit N)	Test: Result less than zero? (Bit 15 = 1)
10	(All)	Load Condition Code Register from Stack. (See Special Operations)
11	(Bit I)	Set when interrupt occurs. If previously set, a Non-Maskable Interrupt is required to exit the wait state.
12	(All)	Set according to the contents of Accumulator A.

LEGEND:

OP	Operation Code (Hexadecimal);
~	Number of MPU Cycles;
#	Number of Program Bytes;
+	Arithmetic Plus;
−	Arithmetic Minus;
•	Boolean AND;
M_{SP}	Contents of memory location pointed to be Stack Pointer;
+	Boolean Inclusive OR;
⊕	Boolean Exclusive OR;
$\overline{M}$	Complement of M;
→	Transfer Into;
0	Bit = Zero;
00	Byte = Zero;

CONDITION CODE SYMBOLS:

H	Half-carry from bit 3;
I	Interrupt mask
N	Negative (sign bit)
Z	Zero (byte)
V	Overflow, 2's complement
C	Carry from bit 7
R	Reset Always
S	Set Always
↕	Test and set if true, cleared otherwise
•	Not Affected

Note – Accumulator addressing mode instructions are included in the column for IMPLIED addressing

Fig. 6.9 How the microprocessor status is saved on the stack

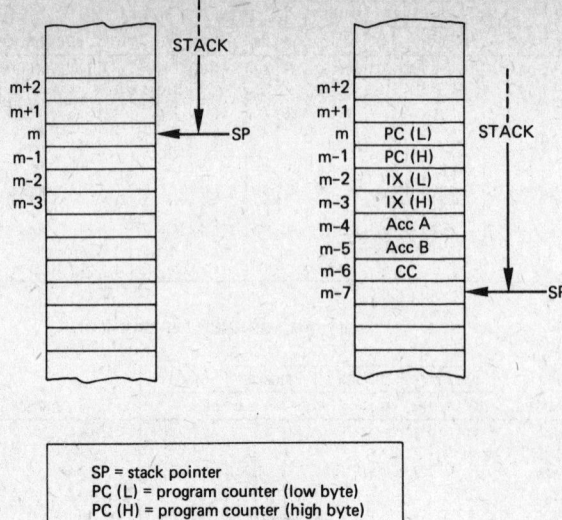

SP = stack pointer
PC (L) = program counter (low byte)
PC (H) = program counter (high byte)
IX (L) = index register (low byte)
IX (H) = index register (high byte)
Acc A = accumulator A
Acc B = accumulator B
CC = condition codes register

Interrupt request $\overline{\text{IRQ}}$ An important control input which allows the processor to be interrupted and for an outside peripheral to be serviced. This occurs if the interrupt bit in the condition code register is clear (i.e. not set). A negative edge on $\overline{\text{IRQ}}$ will then initiate the interrupt. The sequence is as follows:

a The current program instruction will be completed.
b Processor status will be saved on the stack (see *fig. 6.9*).
c The interrupt mask bit in the CCR will be set high so that no further interrupts may be accepted.
d The 16-bit address held at memory locations $FFF8 and $FFF9 will be loaded into the program counter. This address will be the start of the interrupt service routine.
e At the end of the service routine (using instruction RTI), the previous state of the processor registers will be returned from the stack and the main program can continue.

Three-state control (TSC)   This input, when taken high, effectively "disconnects" the processor from the address bus. The TSC input causes the address bus and the R/W̄ line to be put into a high impedance state. At the same time, VMA and BA signals are taken low. This allows the address bus to be used by other devices. Note that a necessary condition is that the ϕ_1 and ϕ_2 clocks must be held high and low respectively.

Data bus enable (DBE) When held low, this pin allows another device to control the data bus. Normally, it is driven by the ϕ_2 clock signal.

Bus available (BA) This output, when in the low state, indicates that the address bus and data bus are being used by the processor. When BA is in the high state, this indicates that the processor has stopped and that the address bus is available. This will occur if the HALT line is in the low state or as a result of a WAIT instruction.

Valid memory address (VMA) This output indicates to peripheral devices that there is a valid address on the address bus. This signal is therefore normally used for enabling peripheral interface adaptors such as the PIA and ACIA.

Read/Write (R/W̄) This TTL-compatible output signals to peripherals and memory chips when the processor is in a Read (high) or Write (low) condition.

6.4 Program Example for the 6800/6802

In order to illustrate some of the instructions and addressing modes of the 6800 the program is given below for the relatively simple task of moving a block of data from one area of memory to another. We shall assume that the block begins originally at address $2000 and consists of 32 numbers (this is 20 in hex.). The block is to be moved to a new start address of $2050. The flowchart for this problem is shown in *fig. 6.10* together with written comments and the addressing mode for each instruction. Inititially the index register is loaded with the hex. number equal to the address of the first byte of data. In the program, the # sign indicates the immediate mode and $ indicates that a hex. number is being used. An arbitrary start address of $1000 is used for the program and no address is specified for the monitor. The program is as follows:

Assembly language				Machine code	
				Address	Code (Hex.)
	LDX	#$2000	Set pointer	1000	CE 2000
LOOP	LDA A	$00,X	Load data	1003	A6 00
	STA A	$50,X	Store data	1005	A7 50
	INX		Increment X	1007	08
	CPX	$2020	Compare	1008	8C 2021
	BNE	LOOP	Branch back	100B	26 F6
	JMP	MON	Return to MON	100D	7E XXXX

Fig. 6.10 Flowchart for the task of moving a block of data

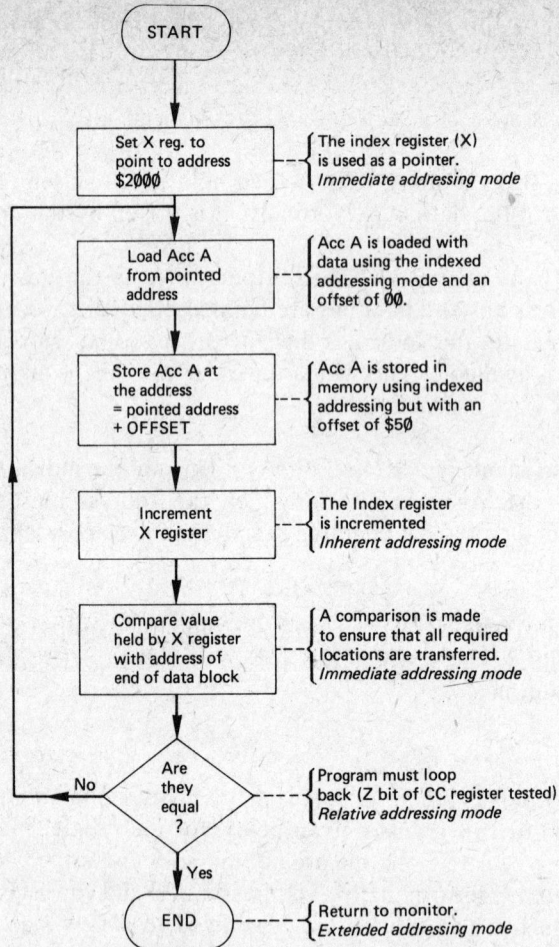

6.5 The PIA type 6821

This is the parallel interface adaptor designed for interfacing between the 6800/6802 processor and peripheral devices. Shown in *fig. 6.11* is a simplified block diagram view of the device. It contains two independent interface circuits (side A and side B) which are identical in most respects except that the B side has a higher drive capability.

Various control lines from the microprocessor to the PIA are necessary:

ENABLE Connected to the ϕ_2 clock line, this is the only timing signal supplied to the PIA.

R/$\overline{\text{W}}$ This input controls whether data is read from (logic 1) or written to (logic $\emptyset$).

Fig. 6.11 MC6821 PIA: simplified block diagram

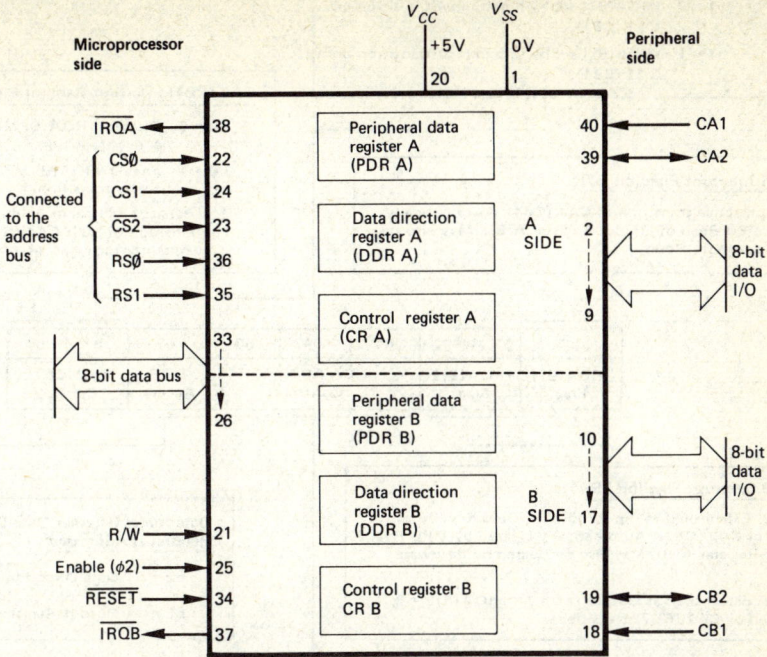

RESET   Used to reset all the internal PIA registers, in other words to clear all registers to zero. This input is usually connected to the reset line of the microprocessor and is therefore active during a reset or power-up condition.

Interrupt Request ($\overline{\text{IRQA}}$ and $\overline{\text{IRQB}}$)   These are outputs from the PIA which are active low to interrupt the microprocessor either directly or through interrupt priority circuits. These outputs are "open drain" type which allows all interrupt request lines to be tied together in a wired-or arrangement.

Each $\overline{\text{IRQ}}$ line from the PIA has two internal flags (these are in the control register, see *fig. 6.12*) which can cause the interrupt request line to go low. These flags are associated with a particular peripheral interrupt line (CA1, CA2, CB1, CB2). The various methods by which the interrupts may be set up and accepted are shown in *fig. 6.12*. The interrupt flags are cleared when the microprocessor executes a read peripheral data register operation.

Before considering the peripheral interface lines, we shall look at the arrangement of the internal registers. Since the A and B sides are identical, the description of operation applies to both sides.

The data lines connected internally to the Peripheral Data Register (PDR) can each be made to act as an input from, or an output to, the peripheral device. The programming to make the lines either inputs or outputs is achieved during initialisation by setting the bits in the Data Direction Register (DDR). This register performs as it is named: that is to say, a logic 1

Fig. 6.12 PIA control register format

Determine Active CA1 (CB1) Transition for Setting Interrupt Flag IRQA(B)1 – (bit b7)

b1 = 0 : IRQA(B)1 set by high-to-low transition on CA1 (CB1).

b1 = 1 : IRQA(B)1 set by low-to-high transition on CA1 (CB1).

CA1 (CB1) Interrupt Request Enable/Disable

b0 = 0 : Disables IRQA(B) MPU Interrupt by CA1 (CB1) active transition.[1]

b0 = 1 : Enable IRQA(B) MPU Interrupt by CA1 (CB1) active transition.

1. IRQA(B) will occur on next (MPU generated) positive transition of b0 if CA1 (CB1) active transition occurred while interrupt was disabled.

IRQA(B) 1 Interrupt Flag (bit b7)

Goes high on active transition of CA1 (CB1); Automatically cleared by MPU Read of Output Register A(B). May also be cleared by hardware Reset.

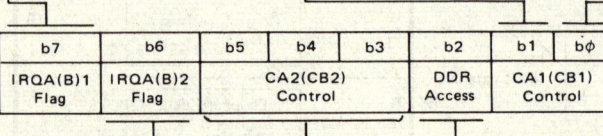

b7	b6	b5	b4	b3	b2	b1	b0
IRQA(B)1 Flag	IRQA(B)2 Flag	CA2(CB2) Control			DDR Access	CA1(CB1) Control	

IRQA(B)2 Interrupt Flag (bit b6)

CA2 (CB2) Established as Input (b5 = 0): Goes high on active transition of CA2 (CB2); Automatically cleared by MPU Read of Output Register A(B). May also be cleared by hardware Reset.

CA2 (CB2) Established as Output (b5 = 1): IRQA(B)2 = 0, not affected by CA2 (CB2) transitions.

Determines Whether Data Direction Register Or Output Register is Addressed

b2 = 0 : Data Direction Register selected.

b2 = 1 : Output Register selected.

CA2 (CB2) Established as Output by b5 = 1

b5	b4	b3
1	0	

(Note that operation of CA2 and CB2 output functions are not identical)

➤ CA2

b3 = 0 : **Read Strobe With CA1 Restore**

CA2 goes low on first high-to-low E transition following an MPU Read of Output Register A; returned high by next active CA1 transition.

b3 = 1 : **Read Strobe with E Restore**

CA2 goes low on first high-to-low E transition following an MPU Read of Output Register A; returned high by next high-to-low E transition.

➤ CB2

b3 = 0 : **Write Strobe With CB1 Restore**

CB2 goes on low on first low-to-high E transition following an MPU Write into Output Register B; returned high by the next active CB1 transition.

b3 = 1 : **Write Strobe With E Restore**

CB2 goes low on first low-to-high E transition following an MPU Write into Output Register B; returned high by the next low-to-high E transition.

b5	b4	b3
1	1	

➤ **Set/Reset CA2 (CB2)**

CA2 (CB2) goes low as MPU writes b3 = 0 into Control Register.

CA2 (CB2) goes high as MPU writes b3 = 1 into Control Register.

CA2 (CB2) Established as Input by b5 = 0

b5	b4	b3
0		

➤ **CA2 (CB2) Interrupt Request Enable/Disable**

b3 = 0 : Disables IRQA(B) MPU Interrupt by CA2 (CB2) active transition.[1]

b3 = 1 : Enables IRQA(B) MPU Interrupt by CA2 (CB2) active transition.

1. IRQA(B) will occur on next (MPU generated) positive transition of b3 if CA2 (CB2) active transition occurred while interrupt was disabled.

➤ **Determines Active CA2 (CB2) Transition for Setting Interrupt Flag IRQA(B)2 – (bit b6)**

b4 = 0 : IRQA(B)2 set by high-to-low transition on CA2 (CB2).

b4 = 1 : IRQA(B)2 set by low-to-high transition on CA2 (CB2).

Fig. 6.13 Setting the state of the peripheral data register

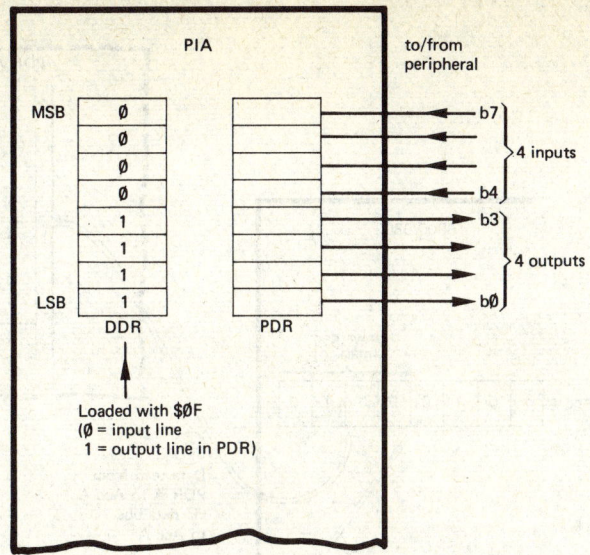

in a bit of the DDR causes the corresponding bit in the PDR to be an output, and a logic ∅ forces it to be an input. This is shown in fig. 6.13 where the word loaded into the DDR is $∅F. This makes b∅ to b3 in the PDR outputs and b4 to b7 inputs. If the DDR was cleared to zeros, all lines to the PDR would be inputs, while if the word loaded was $FF all PDR data lines would be outputs.

In most cases, the PDR will be programmed to act as either all inputs or all outputs, but any combination can be used. If a mix of input/output lines is used, then the read operation from the PDR must be followed by a mask of the lines used as outputs. For example, suppose we have the situation as given in *fig. 6.13*, i.e. b∅ to b3 as outputs and b4 to b7 as inputs. Imagine data loaded into b4 to b7 of the PDR from the peripheral. This could be transferred to Acc A of the 6800/6801 and then b∅ to b3 masked off by an AND instruction (see *fig. 6.14*), i.e.

```
LDA A PDRA      Read PDRA
AND A #$F∅      Mask b∅ to b3
```

The data now left in Acc A will be a four-bit word held in bits 4, 5, 6 and 7.

The control register in the PIA has important functions, for during initialisation the programmer will load the control register, and the control word then sets the way in which the PIA is to operate; in other words, how it is to handle interrupts and in its handshaking. The control register has 8 bits which are used as shown in *fig. 6.15*. Bits b∅ and b1 are used to control the CA1 line; b∅ can be used to mask CA1, because if b∅ is cleared the IRQ line is masked. Bit 1 provides edge control; if b1 = ∅ the interrupt flag b7 responds to the negative (high-to-low) edge of a signal on CA1, whereas if b1 = 1 the IRQ flag is set by the positive (low-to-high) edge.

Fig. 6.14 Masking when a mix of input/output lines is used in the PIA peripheral data register

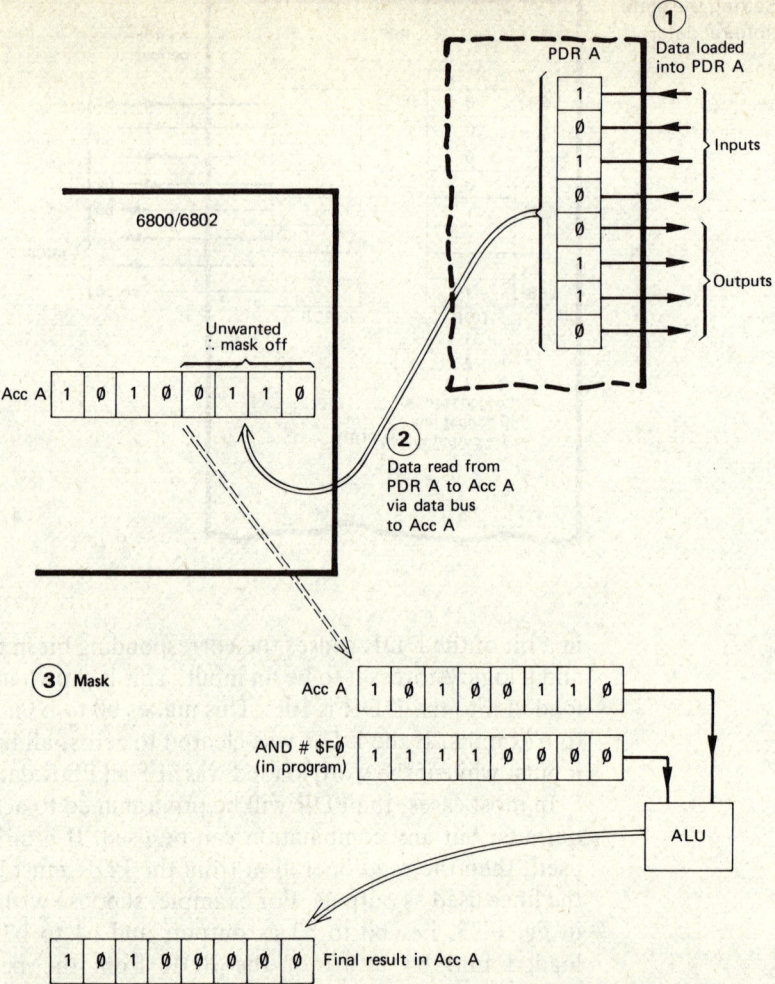

The next bit (b2) in the CR is used to set the addressing for either the Data Direction Register or the Peripheral Data Register. This is necessary because the PDR and DDR share the same address as far as the micro is concerned. If b2 = ∅, the DDR is accessed during addressing and, if b2 = 1, the PDR is accessed during addressing. This is explained more fully in the paragraph on addressing the PIA but essentially the method used during initialisation is, first, to clear the control register (b2 = ∅) and, then when the DDR is to be loaded with the word that sets up either inputs or outputs to the PDR, the common external address for PDR/DDR is selected but internally the data word is routed to the DDR.

Bits 3, 4 and 5 are used to provide CA2 (CB2) control; b5 defines CA2 as an input or output. If b5 = ∅, the CA2 (CB2) is set as an interrupt input. Under these conditions, b3 and b4 are used for masking and edge control respectively. If b5 = 1, CA2 (CB2) is set as an output and b3 and b4 define the mode of operation.

Fig. 6.15a Simple address connections for a PIA

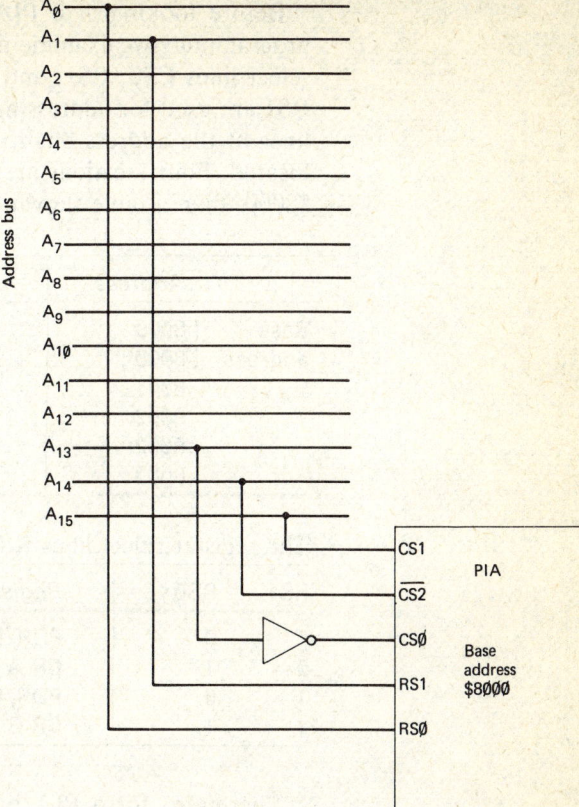

Fig. 6.15b Improved address connections for a PIA (address lines A_{10}, A_{11}, A_{12}, A_{13} are decoded)

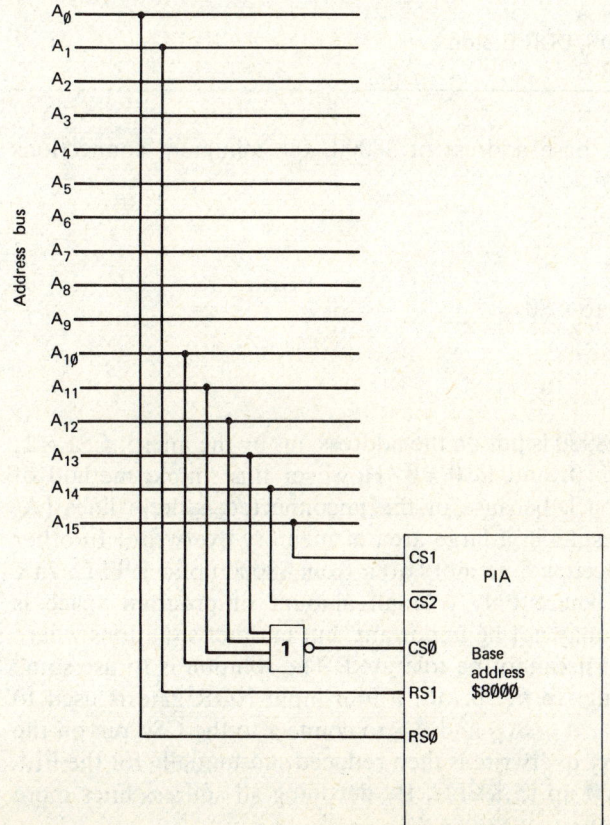

Before looking at a PIA system example in terms of initialisation and programming we examine the way in which the PIA is addressed. Three chip select lines CS∅, CS1, and $\overline{\text{CS2}}$ together with register select lines RS∅ and RS1 are used for addressing. These have to be connected to the appropriate lines of the address bus to determine exactly where in memory the PIA is located. Four locations are required for a PIA. Suppose the base address is $8∅∅∅, then a table showing the various locations is as follows:

	Address	PIA register	Comments
Base address	8∅∅∅	Peripheral data register	A side (bit 2 of CRA 1)
	8∅∅∅	Data direction register	A side (bit 2 of CRA ∅)
	8∅∅1	Control register	A side
	8∅∅2	Peripheral data register	B side (bit 2 of CRB 1)
	8∅∅2	Data direction register	B side (bit 2 of CRB ∅)
	8∅∅3	Control register	B side

The register select lines RS∅ and RS1 are used as follows:

RS1	RS∅	Register selected in PIA
∅	∅	PDR/DDR A side
∅	1	CR A
1	∅	PDR/DDR B side
1	1	CR B

Therefore, for a PIA base address of $8∅∅∅, the minimum connections required are (*fig. 6.15a*):

A_{15} to CS1
A_{14} to $\overline{\text{CS2}}$
A_{13} via an invertor to CS∅
A_1 to RS1
$A_∅$ to RS∅

Then, when address $8∅∅∅ is put on the address bus by the micro, CS1 = 1, $\overline{\text{CS2}}$ = ∅, CS∅ = 1, RS1 = ∅, and RS∅ = ∅. However this simple method of connection has a drawback because of the unconnected address lines (A_2 through to A_{12}). This results in a large area of memory overwrite. In other words, the PIA would occupy a memory area from $8∅∅∅ up to $9FFC. In a control system set up where only a small amount of program space is required, this overwrite may not be important, but in other situations where memory space is limited it cannot be tolerated. The solution is to use some form of address decoding. In *fig. 6.15b*, a four-input NOR gate is used to decode address lines A_{10}, A_{11}, A_{12} and A_{13} to connect to the CS∅ pin on the PIA. The area of memory overwrite is then reduced substantially for the PIA now occupies from $8∅∅∅ up to $83FC. By decoding all address lines there will be no overwrite at all.

Fig. 6.16 Initialisation for the PIA

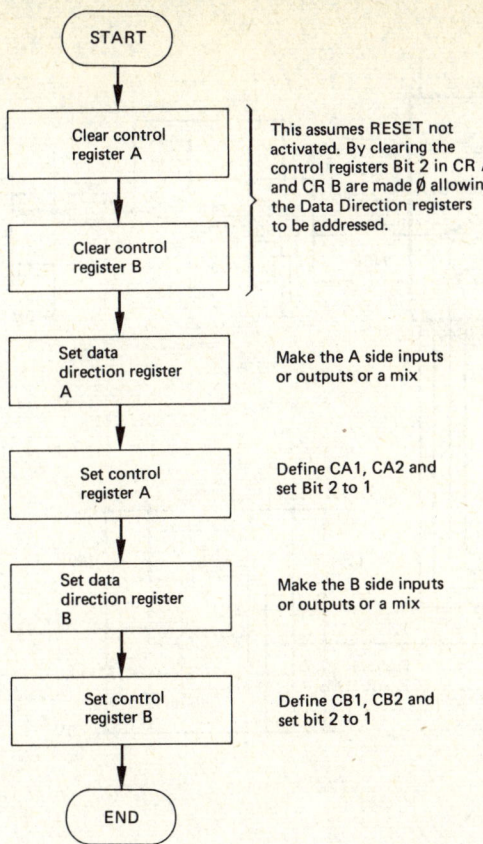

The method by which the PIA is set up or configured to suit a particular system requirement is called *initialisation*. A flowchart showing the steps for initialisation is given in *fig. 6.16*. The next section will illustrate the technique.

6.6 A Simple PIA Application

Using a 6802D5E evaluation kit, a small d.c. permanent magnet motor is to have its speed set and controlled (*fig. 6.17*). The circuit for the motor drive was given in *fig. 4.30* (please refer back to page 98). On the D5 kit, the user PIA is at address $E48\emptyset$ to $E483 and 4 bits of the PDR on the B side are used for the control. These must be programmed as outputs. A further requirement is that an external switch is allowed to interrupt the program which then causes the motor to stop and remain off for approximately 20 seconds. When the external push switch is operated, the 555 timer, wired as a monostable, will trigger on, and the leading edge of the pulse will initiate the interrupt. A monostable is used to ensure that the switch signal is fully debounced (see next chapter for other methods of switch debouncing). Assume that the main program has the task of running the motor up from the off condition to full speed in say 15 seconds, and then to leave it running at full speed for 30 seconds (note that this is only an example).

148 Practical Interface Circuits for Micros

Fig. 6.17 A motor control interface using the PIA

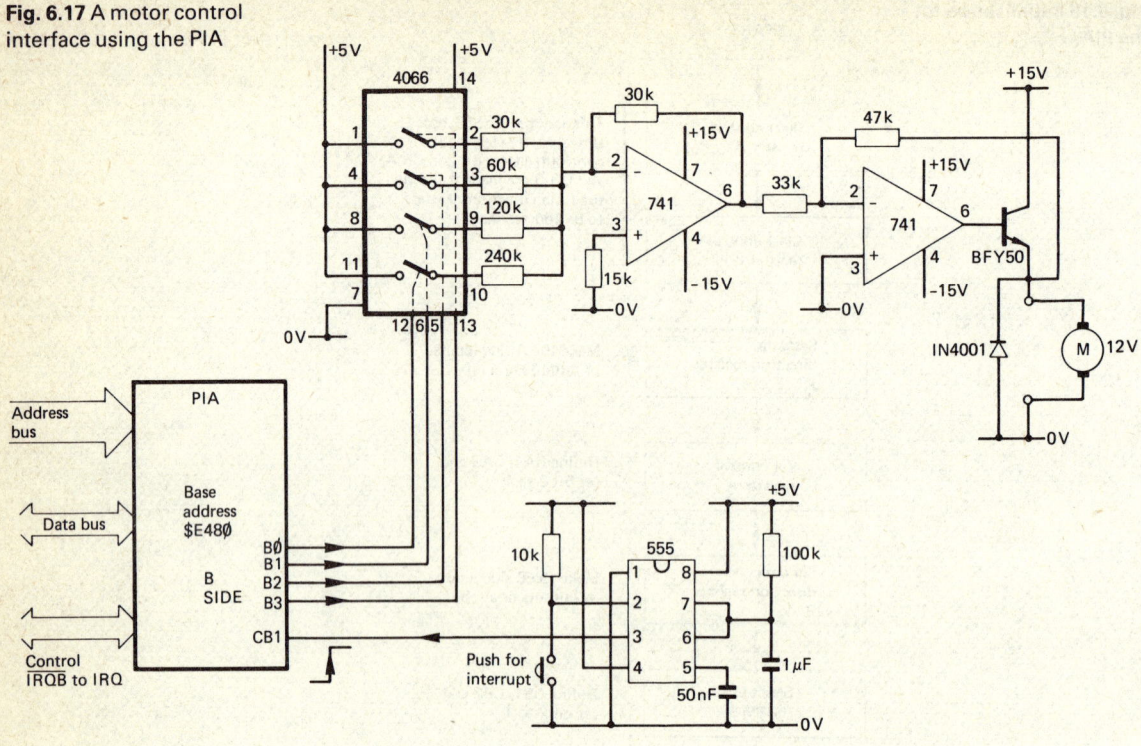

There must be four sections of program placed in memory to carry out the required task (for flowcharts see *fig. 6.18*); we shall assume that the interrupt mask bit is clear:

1) Initialisation of the PIA.
2) Setting the stack and interrupt pointers.
3) Speed control program (with a time delay subroutine).
4) The interrupt request program.

The user RAM (inside the 6802) is from $\$0000$ to $\$007F$ and the user IRQ vector is at address $\$E43C$. The start address of the interrupt request program must be stored in this vector. It is also important that the stack pointer is set correctly so that, when an interrupt occurs, the status (i.e. contents of the internal registers) is saved. We shall set the pointer to the top of the available RAM in the 6802, i.e. at address $\$007F$.

Fig. 6.18 Flowcharts for PIA application

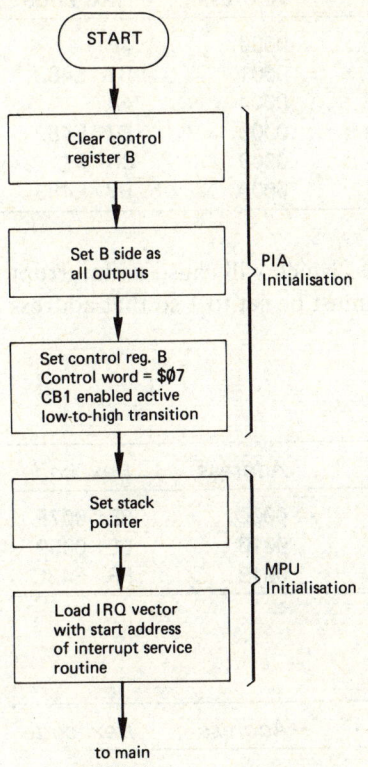
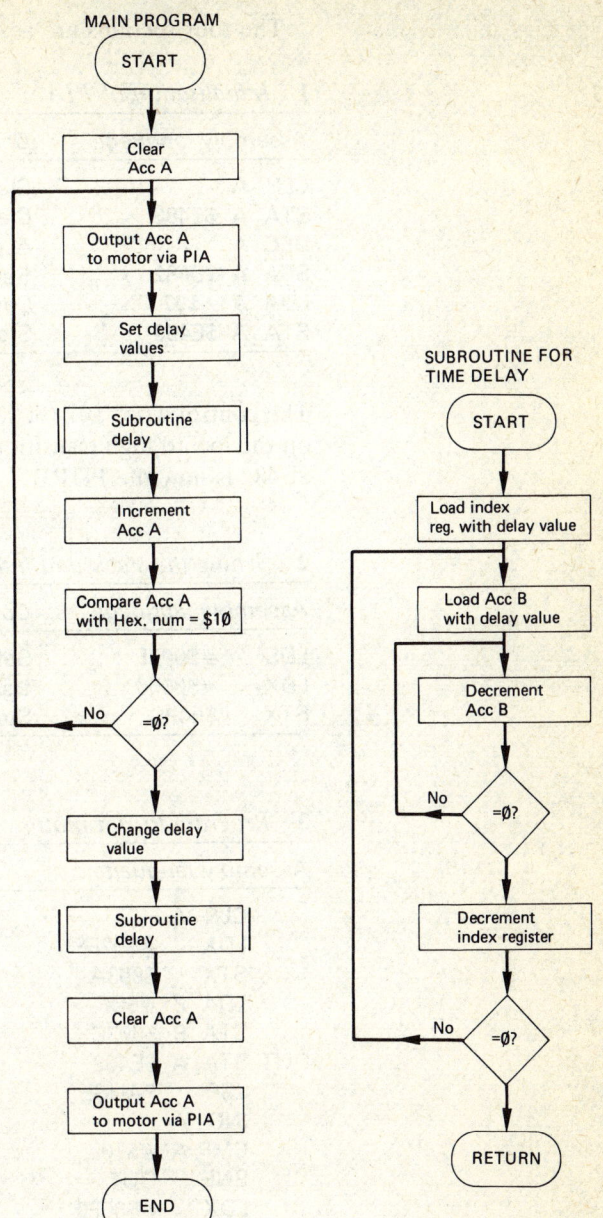

The four sections are as follows:

1 Initialisation of PIA

Assembly language	Comments	Address	Hex. code
CLR A	Clear Acc A	ØØØØ	4F
STA A $E483	Clear CRB	ØØØ1	B7 E483
DEC A	Acc A now holds $FF	ØØØ4	4A
STA A $E482	Set PDRB as outputs	ØØØ5	B7 E482
LDA A #$Ø7	Control word	ØØØ8	86 Ø7
STA A $E483	Store in CRB	ØØØA	B7 E483

The control word $Ø7 (ØØØØØ111) enables CB1 which will cause an interrupt on the low to high transition. Note that bit 2 must be set to 1 so that address $E482 is now the PDRB.

2 Setting the stack and interrupt pointers

Assembly language	Comments	Address	Hex. code
LDS #$ØØ7F	Set stack pointer	ØØØD	8E ØØ7F
LDX #$ØØ5Ø	User IRQ address	ØØ1Ø	CE ØØ5Ø
STX $E43C	Store in vector	ØØ13	FF E43C

3 Program to run motor

Assembly language	Comments	Address	Hex. code
CLR A	Clear Acc A	ØØ16	4F
LDX #$Ø2FF	Delay value 1	ØØ17	CE Ø2FF
STX $ØØ3A	Store value 1	ØØ1A	DF 3A
LDA B #$CF	Delay value 2	ØØ1C	C6 CF
STA B $ØØ3C	Store value 2	ØØ1E	D7 3C
OUT STA A $E482	Set motor	ØØ2Ø	B7 E482
JSR PAUSE	Sub delay	ØØ23	BD ØØ3D
INC A		ØØ26	4C
CMP A #$1Ø	Max speed?	ØØ27	81 1Ø
BNE OUT	Branch back	ØØ29	26 F5
LDX #$66FØ	Delay value 3	ØØ2B	CE 66FØ
STX $ØØ3A	Store value 3	ØØ2E	DF 3A
JSR PAUSE	Sub delay	ØØ3Ø	BD ØØ3D
CLR A	Clear Acc A	ØØ33	4F
STA A $E482	Motor off	ØØ34	B7 E482
JMP MON	End	ØØ37	7E FØ24

Within this program a set of delay values used in the Delay Subroutine are set up and stored. The addresses used for the delay values are $ØØ3A (double byte) and $ØØ3C. The Delay Subroutine is as follows:

Assembly language			Comments	Address	Hex. code	
PAUSE	LDX	$003A	Fetch delay value	003D	DE	3A
LOOP 2	LDA B	$003C	Fetch delay value	003F	D6	3C
LOOP 1	DEC B		⎫	0041	5A	
	BNE	LOOP 1	⎬ Time delay	0042	26	FD
	DEX		⎬	0044	09	
	BNE	LOOP 2	⎭	0045	26	F8
	RTS			0047	39	

4 Interrupt request program

Assembly language			Comments	Address	Hex. code	
	LDA A	$E482	Get port value	0050	B6	E482
	PSH A		Store port value	0053	36	
	CLR A		Clear Acc A	0054	4F	
	STA A	$E482	Motor off	0055	B7	E482
	LDX	#$2F00		0057	CE	2F00
LOOP4	LDA B	#$FF	⎫	005A	C6	FF
LOOP3	DEC B		⎬ Time delay	005C	5A	
	BNE	LOOP3	⎬	005D	26	FD
	DEX		⎬	005F	09	
	BNE	LOOP4	⎭	0060	26	F8
	TST	$E482	Clear flag	0062	7D	E482
	PUL A		Restore port	0065	32	
	STA A	$E482		0066	B7	E482
	RT1		Return	0069	3B	

This portion of the program will stop the motor and hold it off for about 20 seconds. To clear the IRQB flag in CRB of the PIA, it is necessary to read or test the B data register (PDRB); this is the last instruction before RTI, the Return from Interrupt.

6.7 The Asynchronous Communications Interface Adaptor (ACIA) M6850

This is an interface adaptor which provides controlled connection to and from the 6800 series microprocessors and peripherals that receive and transmit serial data. The 8-bit parallel data from the micro is stored in a data register within the ACIA, it is formatted (i.e. provided with start, stop and parity bits), and then outputted one bit at a time at a rate acceptable to the peripheral. Communication is, of course, possible the other way round, with serial data from the peripheral being converted by the ACIA into parallel form suitable for the microprocessor. Thus, the main uses of an ACIA are in interfacing between the microprocessor and equipment, such as printers, teletypes, and cassette tape, and in transmitting and receiving data over a telephone line or radio link. In this case a modulator/demodulator IC called a modem would be required. A **modem** converts the serial data into audio tones and vice versa. A logic 1 is represented by a burst of pulses at a frequency of 2400 Hz and a logic 0 by a burst of pulses at 1200 Hz.

Fig. 6.19 Format used for asynchronous serial data

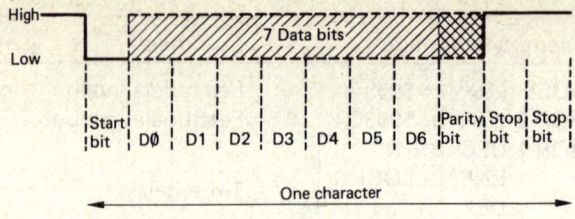

Before studying the internal structure and operation of the ACIA, we look at the format used for asynchronous serial data transmission (*fig. 6.19*). The signal line is normally high so the beginning of a character is indicated by a start bit going low. The data bits, seven for ASCII characters, are then set and a parity bit (D7) can be used for error detection. If, for example, even parity is used, selected during the initialisation routine of the ACIA, then the parity bit is set to $\emptyset$ or to 1 to make the sum of bits D$\emptyset$ to D7 an even number. Errors during transmission will be indicated by a logic 1 in bit 6 of the ACIA status register. Following the parity bit, the end of the character is indicated by two stop bits, both high states.

The speed at which bits are sent, the number of bits per second, is called the **baud rate**. Typical baud rate values are 110, 150, 300, 600 and 1200 for microprocessor systems. This value will include transmission of the start, parity and stop bits which means that the rate of data transmission is lower than the baud rate. If the baud rate is 110, then the rate of actual data transfer is 80 data bits per second.

Fig. 6.20 Simplified view of an ACIA

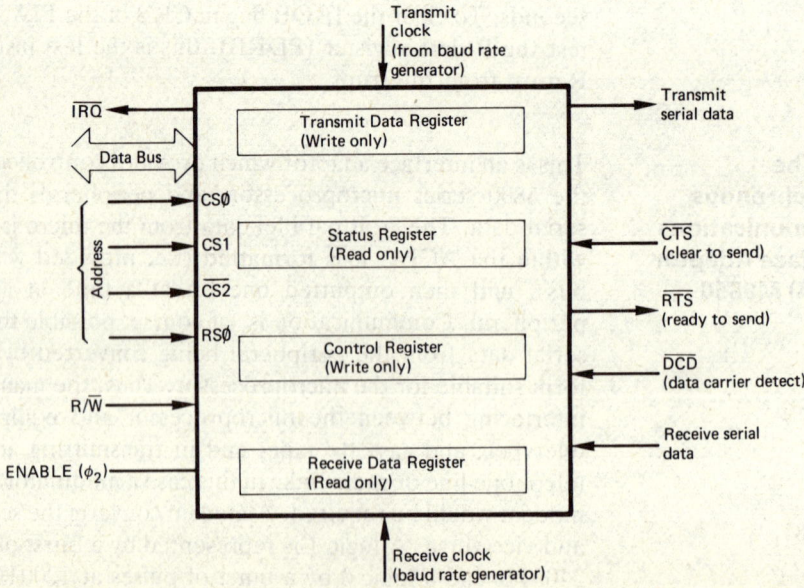

The simplified view of the ACIA in *fig. 6.20* shows that it contains four main registers:

1) Transmit Data Register TDR (Write only)
2) Status Register SR (Read only)
3) Control Register CR (Write only)
4) Receive Data Register RDR (Read only)

Since a register is either a write only or read only, the read/write line is used as part of the addressing. The ACIA is memory-mapped using $CS0$, CS1, $\overline{CS2}$ and RS0 and it appears as only two addressable memory locations. The register select pin together with $R/\overline{W}$ is used to select each register as shown in the table:

Address line A0 connected to RS0	Read/Write $R/\overline{W}$	Register selected in ACIA
0	0	CR
0	1	SR
1	0	TDR
1	1	RDR

The device requires an external transmit and receive clock circuit but the clock can be divided down by 1, 16 or 64 by setting the first two bits in the control register:

Bit 1	Bit 0	Result
0	0	÷1 clock
0	1	÷16 clock
1	0	÷64 clock
1	1	Master reset

For interfacing with modems, three signals, which permit handshake facility, are provided:

Clear to Send	$\overline{CTS}$
DTA Carrier Detect	$\overline{DCD}$
Request to Send	$\overline{RTS}$

When a modem is not used, $\overline{CTS}$ and $\overline{DCD}$ are connected to 0 V. $\overline{RTS}$ can be left unconnected.

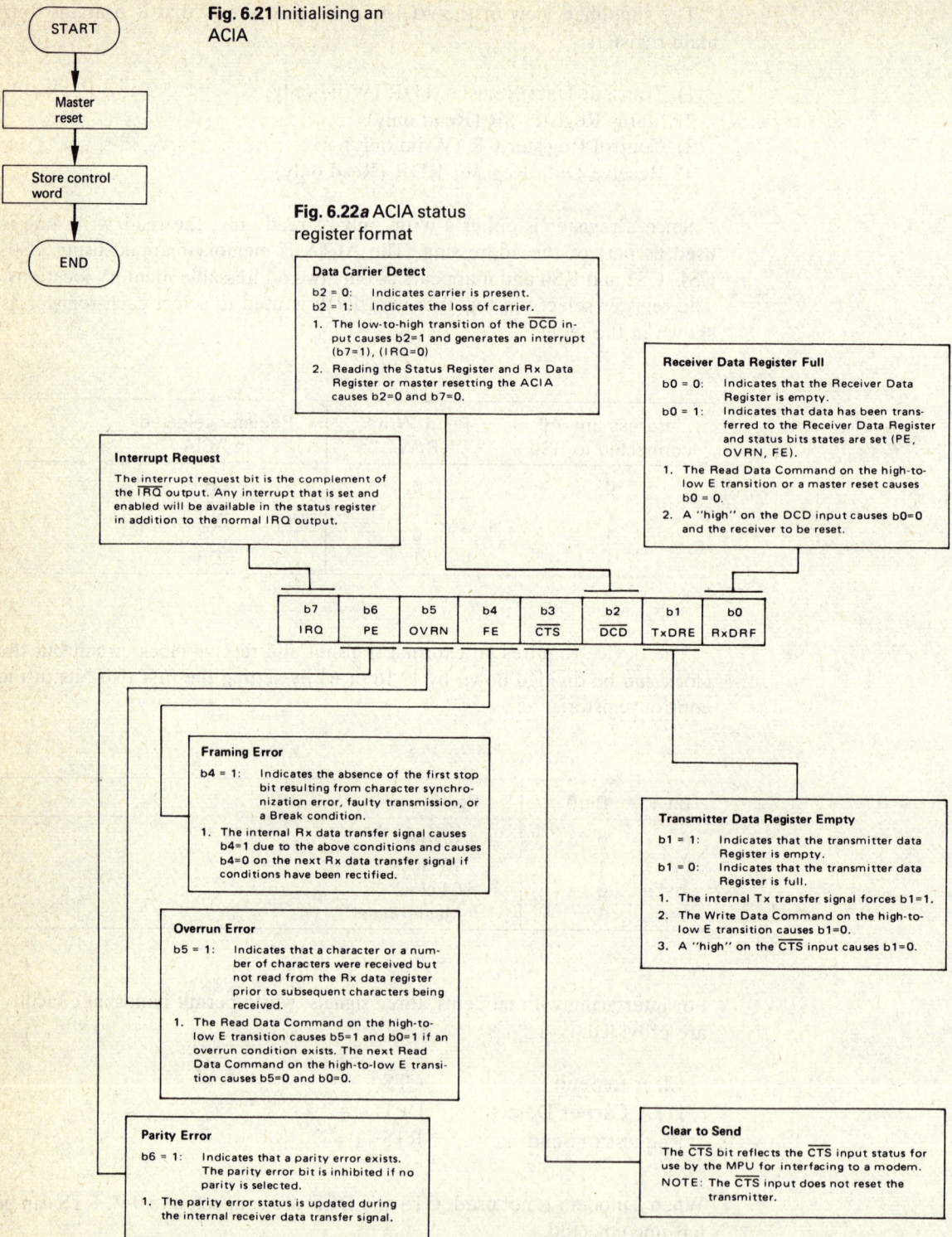

The initialisation of the ACIA is shown in flowchart form in *fig. 6.21*. A typical assembly language initialisation for an ACIA at base address D∅∅∅ could be

LDA	A	#$∅3	} Master Reset
STA	A	$D∅∅∅	Clears all ACIA Register
LDA	A	#$∅5	} Control word gives 7 data bits/odd parity/stop bits
STA	A	$D∅∅∅	and a ÷16 clock.

The formats for the control and status registers are shown in *figs. 6.22a and b*.

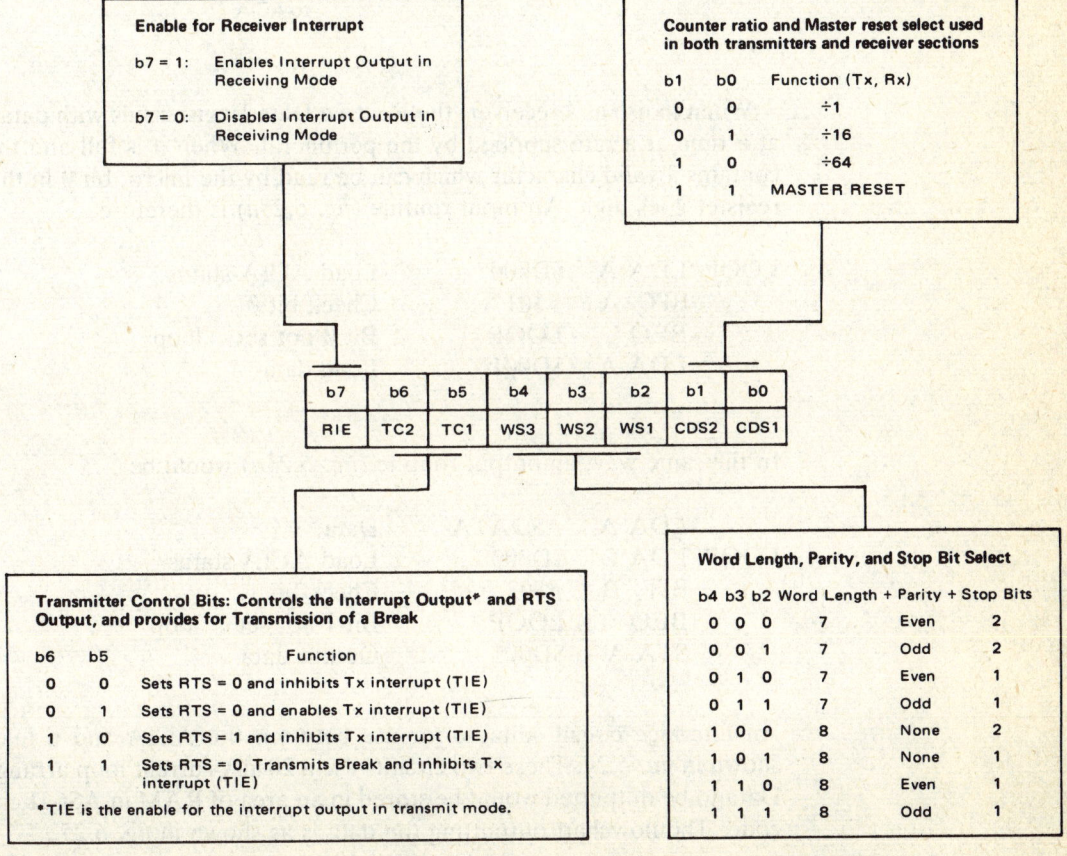

Fig. 6.22b ACIA control register format

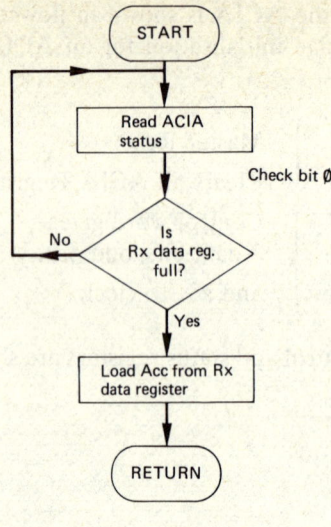
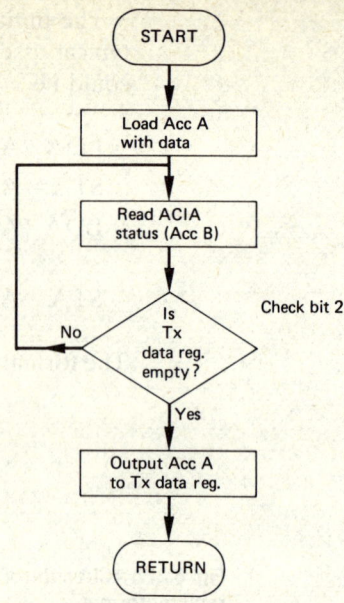

Fig. 6.23a Input routine from ACIA

Fig. 6.23b Output routine to ACIA

When in use as a receiver, the Receive Data Register fills with data one bit at a time at a rate supplied by the peripheral. When it is full and therefore contains a valid character which can be read by the micro, bit ∅ in the status register goes high. An input routine (*fig. 6.23a*) is therefore

```
LOOP  LDA A   $D∅∅∅        Load ACIA status
      BIT A   #$∅1         Check bit ∅
      BEQ     LOOP         Bit ∅ not set ∴ loop
      LDA A   $D∅∅1        Load data
```

In the same way, an output routine (*fig. 6.23b*) would be

```
      LDA A   #$DATA       Data
LOOP  LDA B   $D∅∅∅        Load ACIA status
      BIT B   #$∅2         Check bit 1
      BEQ     LOOP         Bit 1 not set ∴ loop
      STA A   $D∅∅1        Output data
```

An interface circuit suitable for use between the ACIA and a teletype is shown in *fig. 6.24*. These two circuits use a 20 mA current loop arrangement. Data to be outputted would be stored in an area of RAM in ASCII character code. The flowchart outputting the data is as shown in *fig. 6.25*.

Microprocessors and Interface Adaptors 157

Fig. 6.24 ACIA to teletype interface

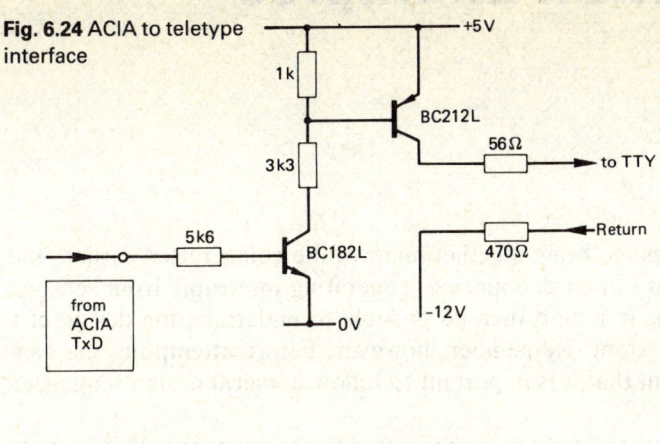

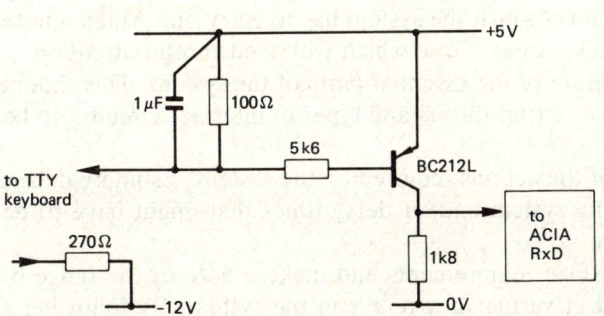

Fig. 6.25 Flowchart for outputting data using ACIA

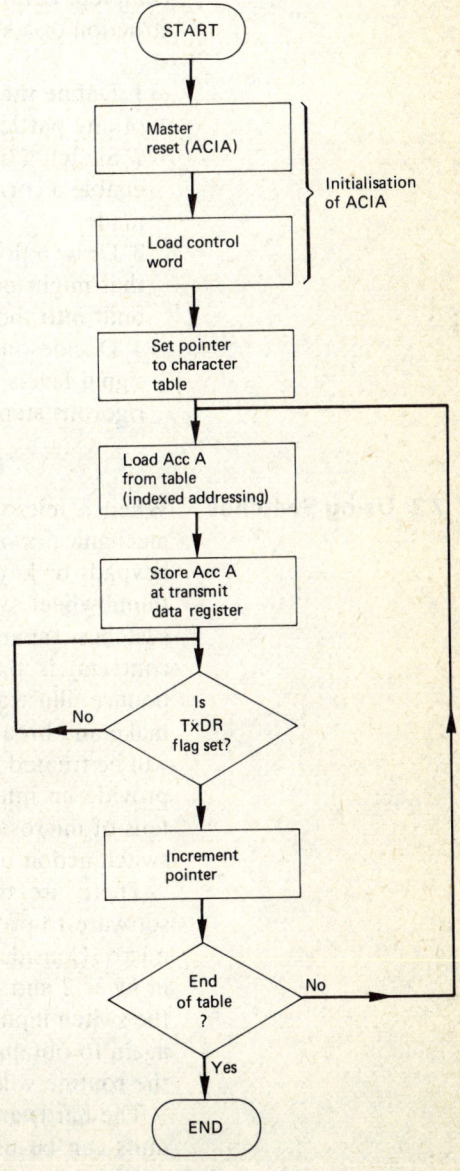

7 Application Examples

7.1 Introduction

This chapter attempts to bring together many of the points raised earlier, and in addition looks at switch debouncing, generating interrupts from sensors, and display driving. It should then be possible to undertake the design of a complete control system. Remember, however, before attempting the construction of a system that it is important to follow a logical design sequence:

1 Define the exact task(s) which the system has to carry out. Make a note of any particularly tricky areas, those which will need careful attention.
2 Sketch a block diagram of the essential parts of the system. This should enable a correct choice of transducers and types of interface circuitry to be made.
3 Draw a flowchart of the actions required by the system. Estimate delays that might occur in the system and/or delay times that might have to be built into the program.
4 Decide on the interface requirements and make a note of the range of signal levels expected at various key points in the system. By following a rigorous step-by-step sequence, many of the pitfalls may be avoided.

7.2 Using Switches

When a micro is used for control, many of its input signals will be from mechanical switches or contacts. These will be single push-button switches, keypads or keyboards, reed switches used for limit and position sensing, or thumbwheel switches for setting-up purposes. The problem with all these switches (apart from special types such as those with mercury wetted contacts) is that they are always subject to severe contact bounce. This bounce, illustrated in *fig. 7.1*, can take up several milliseconds as the contacts make and break before finally settling to the new position. All the variations will be treated as valid inputs to the system. Suppose a push switch is used to provide an interrupt as shown. The interrupt service routine may only last tens of microseconds and would therefore be triggered several times by one switch action unless the contact bounce is eliminated.

There are two possible solutions to this switch problem, to use either software to produce a delay or hardware to debounce the input edge to the micro. Consider the software approach first. The flowchart required is shown in *fig. 7.2* and basically consists of introducing a delay of, say, 20 msec after the switch input has first been sensed and then reading the state of the switch again to obtain the steady state condition. If the switch remains closed then the routine will be carried out; if it is open the routine is exitted.

The hardware solution can take many forms. Special "debounced" switch units can be used; these will either be Hall-effect devices or contain some

Application Examples 159

Fig. 7.1 Contact bounce in system using a mechanical switch

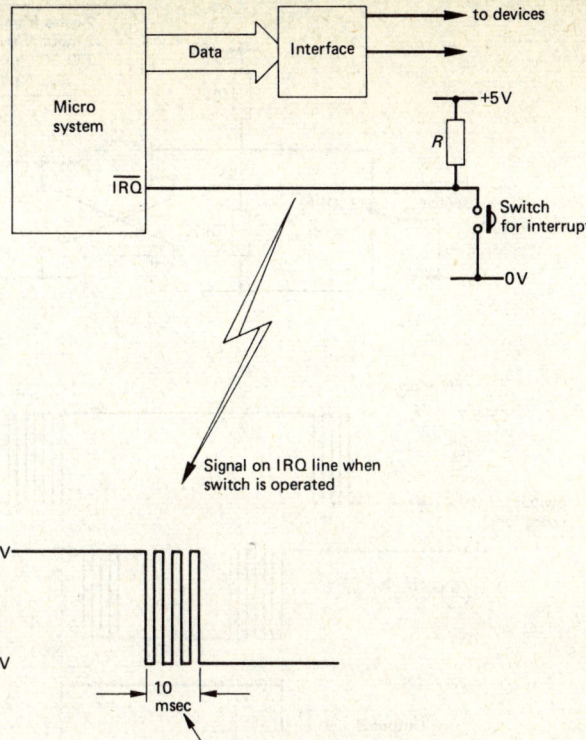

Fig. 7.2 Software solution to switch bounce

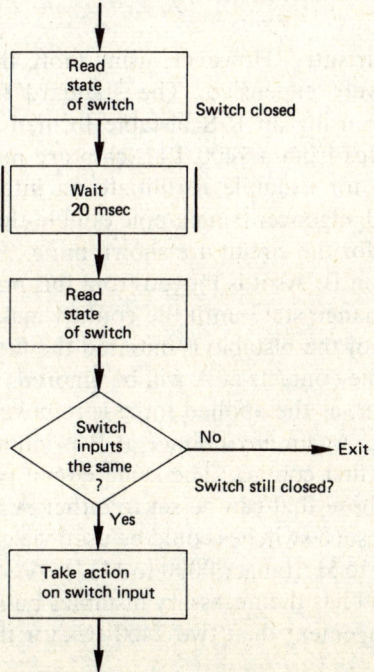

Fig. 7.3 Hardware solution to switch bounce using R-S bistable circuit

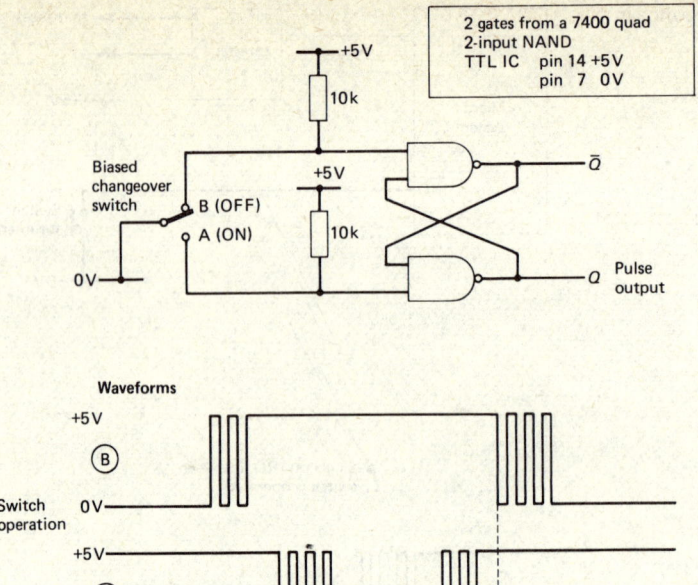

built-in debounce circuitry. However, using more than a few of these types could prove relatively expensive. The standard hardware solution is to interface the switch using an R-S bistable formed from two cross-coupled NAND gates (2 gates from a 7400 TTL chip are normally used). If a pulse output is required, for example to initiate an interrupt, then a biased or momentary action changeover (single-pole double-throw) switch is necessary.

The connections for the circuit are shown in *fig. 7.3*. Assume the switch is biased OFF at position B. As it is moved from this position, the output of the bistable does not change state until the contact makes with position A. But the change of state of the bistable is initiated the first time A goes to logic ∅ and the bounce of the contacts at A will be ignored. When the switch returns to its biased position, as the applied force is removed, B goes to logic ∅ and the bistable is reset. Again any bounce at B is ignored because the bistable will respond to the first contact. The same circuit can be used to debounce latched switches—those that can be set to either A or B and will remain set until changed. Four such switches could be used via an input port to set a hex. input value from $∅ to $F (binary ∅∅∅∅ to 1111). A contact bounce eliminator chip such as the 8544 has the necessary bistables built in and might provide a more suitable arrangement than two 7400 ICs for the debounce of the four switches (see *fig. 7.4*).

Application Examples

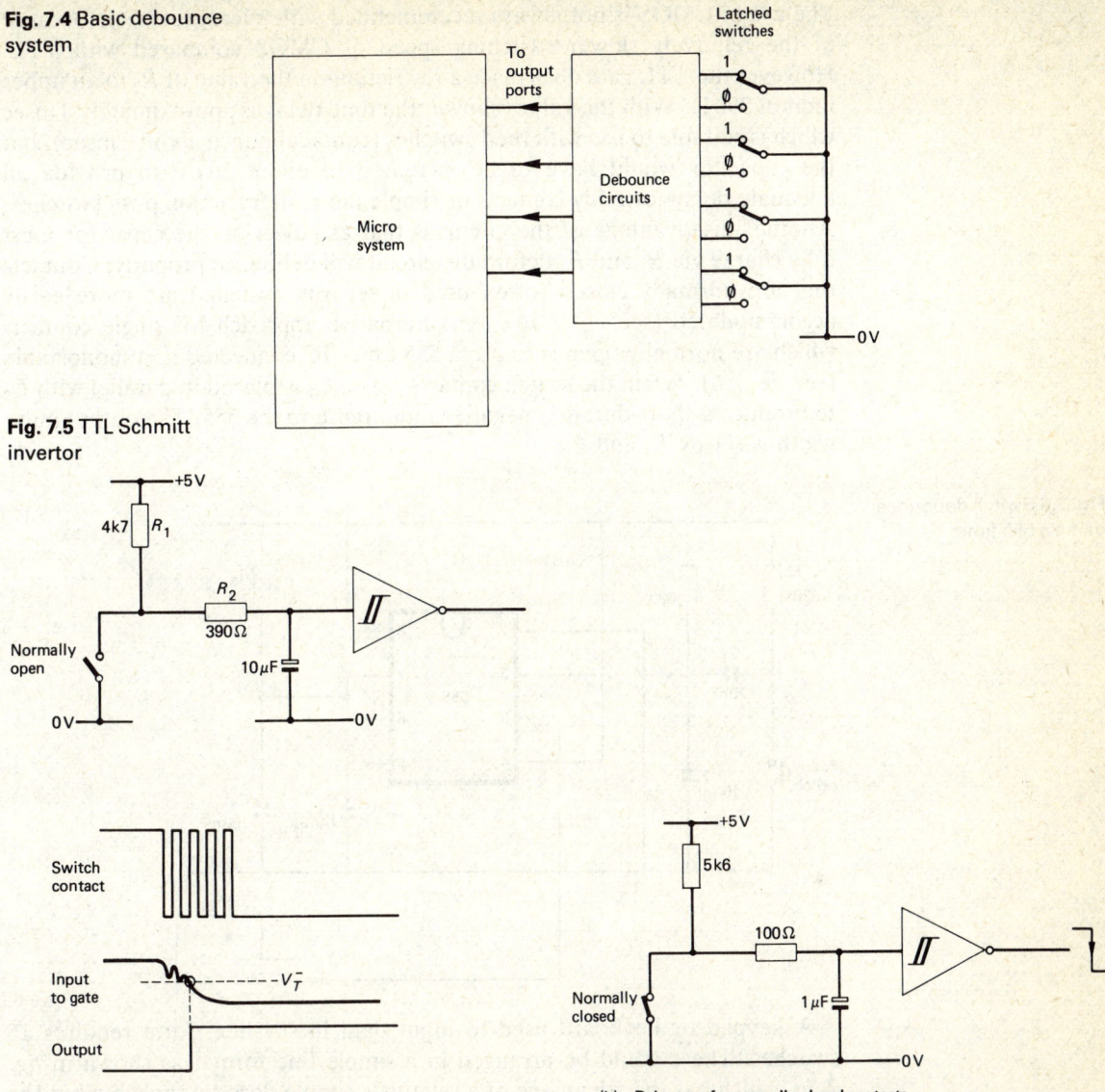

Fig. 7.4 Basic debounce system

Fig. 7.5 TTL Schmitt invertor

(a) Debounce for normally open switch

(b) Debounce for normally closed contacts

In many situations a switch with changeover contacts is not available to provide an input. A reset or interrupt switch needs only to "make" from a normally open position, and limit or position sensing switches will normally consist of a single pair of contacts. If an edge signal is required from such a switch then an electronic delay must be built in to eliminate the bounce. The circuit in *fig. 7.5a* illustrates the principle. When the contacts are made, R_2 discharges C and the TTL Schmitt gate output only changes state when the voltage across C has fallen just below the negative going threshold (V_T^-). Although longer delays could more easily be achieved using a CMOS Schmitt, because of its very high input resistance, this is not really advisable.

The use of CMOS is not always recommended with microprocessors because of the relatively slower switching speed of CMOS compared with TTL. However the TTL gate does place a restriction on the value of R_2 to an upper limit of 390 Ω. With the values shown, the time delay is approximately 4 msec which is suitable to use with reed switches (contact bounce about 2 msec), but the capacitor would have to be increased to about 50 μF to provide an adequate delay for relay contacts or simple momentary action push switches. Another disadvantage of the circuit is that at power-on the capacitor must fully charge via R_1 and R_2 before the circuit will debounce properly. Contacts that are normally closed (often used in security systems) are more easily accommodated (see *fig. 7.5b*). An alternative approach for single contacts which are normally open is to use a 555 timer IC connected as a monostable (see *fig. 7.6*). When the switch contacts close, C_1 is placed in parallel with C_2 to produce a short-duration negative input pulse to the 555. The output pulse width is set by R_3 and C_3.

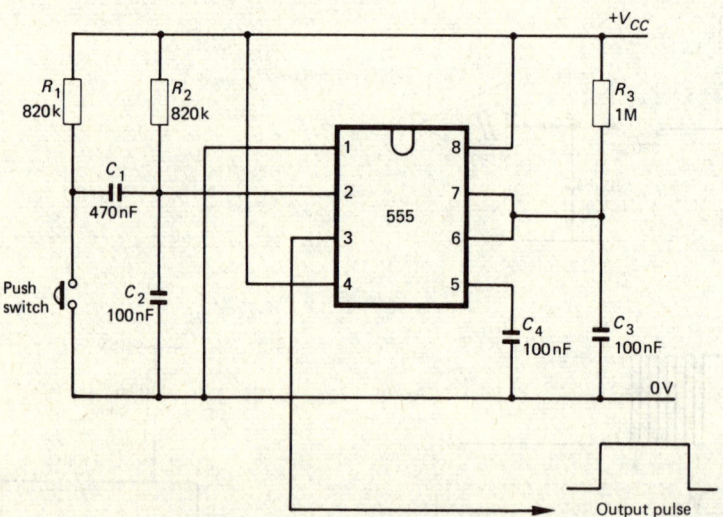

Fig. 7.6 Switch debounce using a 555 timer

A keypad or keyboard used to input data in say hex. form requires 16 switches. These could be arranged in a simple line format as shown in *fig. 7.7a*, which has the advantage of a relatively simple decoding scheme but the disadvantage that a large number of input connections are required. The usual method is to use a matrix-type keypad arrangement to input hex. data (*fig. 7.7b*). Here the switches are arranged in a 4 × 4 array. The software for detecting which key has been depressed is more complicated but only 8 input lines are required.

Before looking at the software, consider the problem of detecting any key depression. One method could be to perform software polling of the keypad. The method is to scan the keys regularly, say at 25 times a second, to pick up a key depression. The time between polls is then available for the micro to carry out other program tasks. If this time loss presents a problem then an interrupt driven system is the alternative. When a key is depressed, an

Application Examples 163

Fig. 7.7a Simple line format for switch input

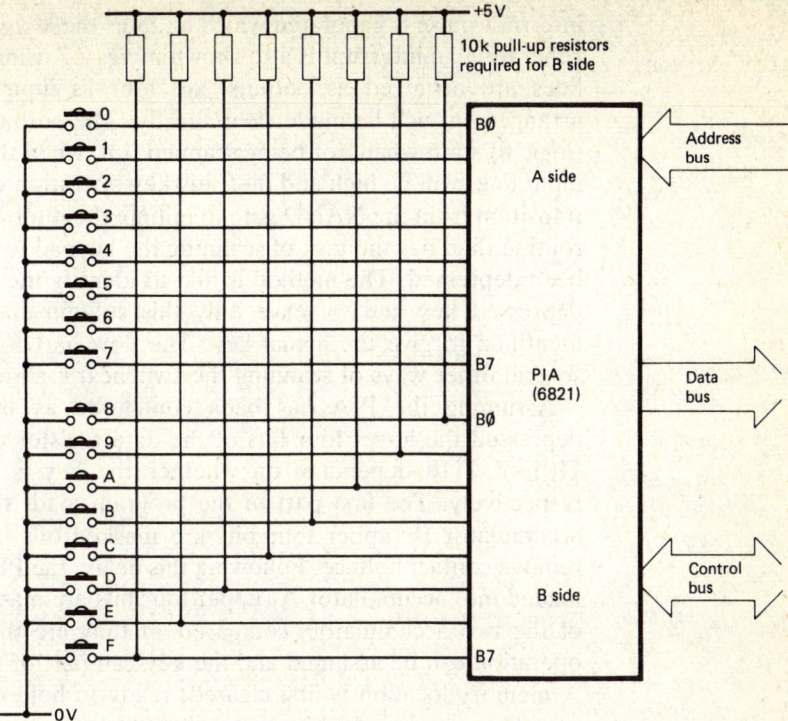

Fig. 7.7b Matrix arrangement for keypad

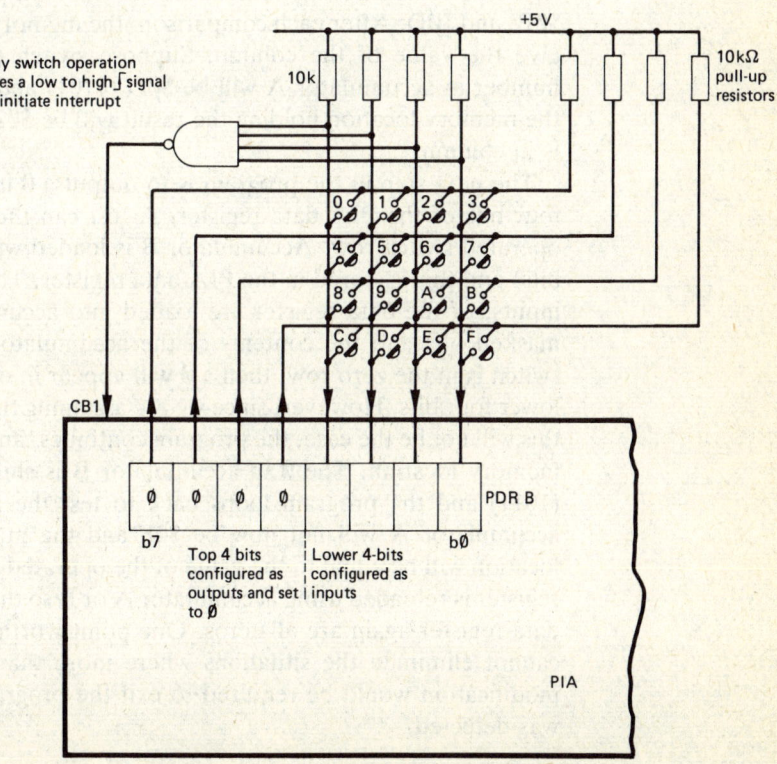

interrupt pulse is generated which initiates the keyscan routine. A system for generating an interrupt is also shown in *fig. 7.7* using a 6821 PIA. Four of the lines are arranged as outputs and four as inputs. The reasons for this arrangement will be made clear shortly. The output lines are left in the low (logic ∅) state when not being scanned, i.e. when the keypad is unused. The input lines will be high and then any key operation will produce a low to high transition from the NAND gate to initiate the interrupt. The interrupt service routine then has the task of scanning the keypad to determine which key has been depressed. The method is first to identify the column connected to the depressed key and to leave only this column energised while the row is identified to give the actual key. The flowchart for one method (there are several other ways of scanning the switches) is shown in *fig. 7.8*.

Assuming the PIA has been configured as shown, when one key is depressed the lower four bits of the data register will be either ∅111, 1∅11, 11∅1 or 111∅ depending on whether the key is in column ∅, 1, 2 or 3 respectively. The first part of the program loads these lower four bits into accumulator B (upper four bits are masked off), then waits say 5 msec to remove contact bounce. Following this delay, the PIA data register content is loaded into accumulator A (upper four bits are masked off) and the contents of the two accumulators compared. If they are the same then a valid key operation can be assumed and the keyscan part of the program commences. A memory location is first cleared, ready to hold the resulting value of the key. The contents of accumulator A are then compared successively with $∅7, $∅B, and $∅D. After each comparison, the memory location is incremented to give the value of the column. Suppose switch 6 has been operated. The number in accumulator A will be $∅D (11∅1) and after the $∅D comparison the memory location holding the result will be $∅2, indicating that the switch is in column 2.

The next step in the program is to output a ∅ in turn to each of the upper four bits of the PIA data register. A test can then be made to indicate the operated switch row. Accumulator B is loaded with $7∅ (∅111 in upper four bits) and this is stored in the PIA data register. The lower four bits, acting as inputs, of the data register are loaded into accumulator A (upper four bits masked off) and the contents of the accumulator compared to $∅F. If the switch is in the zero row, then a ∅ will appear in one bit of the accumulator's lower four bits. However, since we are assuming that key switch 6 is operated, this will not be the case, the program continues, and a $∅4 will be added to the memory location. The ∅ in accumulator B is shifted one place to the right (1∅11) and the program loops back to test the next row. The contents of accumulator A will not now be $∅F and the number held in the memory location will be equal to the value of the operated key ($∅6). Finally, the data register is reloaded using accumulator A or B so that the upper four bits of the data register again are all zeros. One point worth noting is that this method cannot eliminate the situations where more than one key is depressed. A modification would be required to exit the program if an invalid input code was detected.

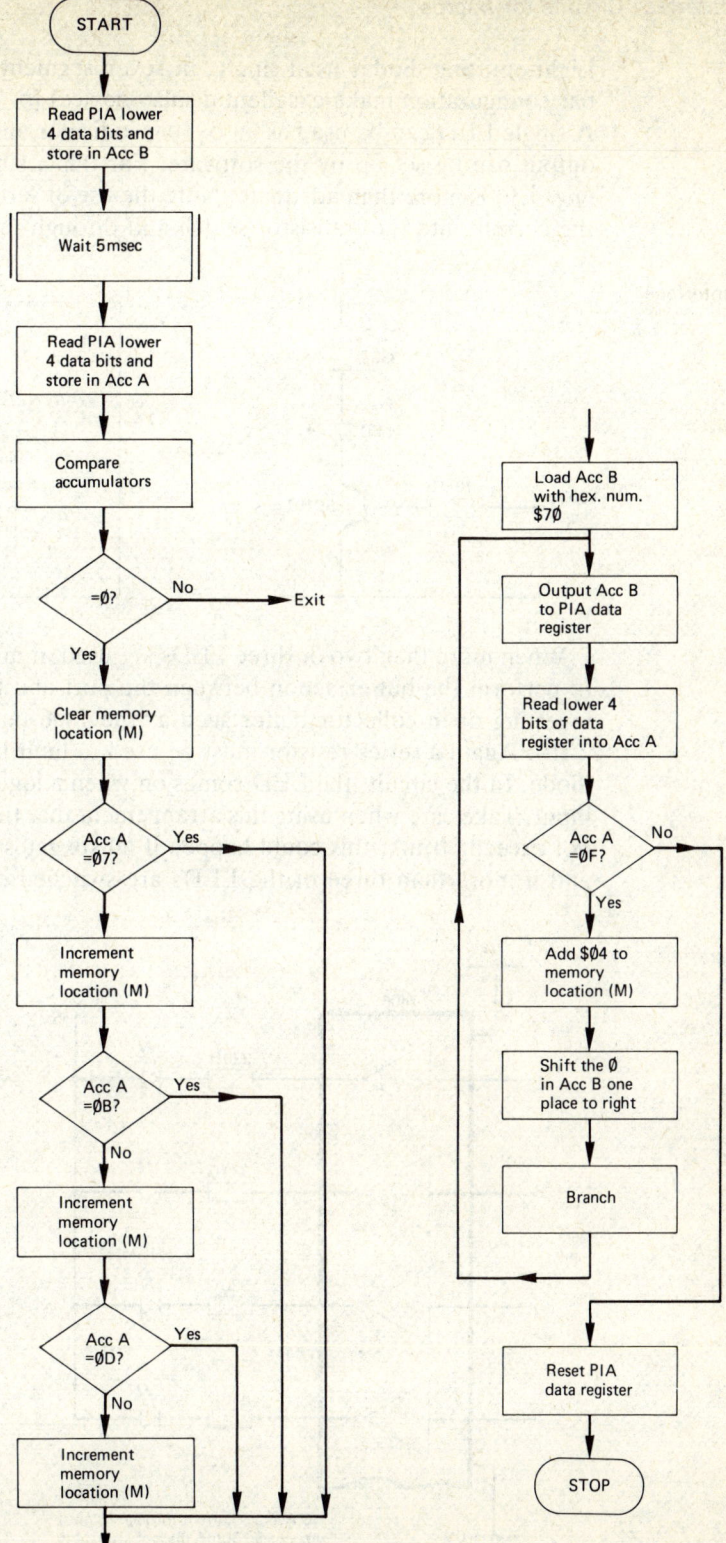

Fig. 7.8 Flowchart for keypad decode routine

7.3 Driving Displays

Light-emitting diodes used singly, in seven-segment units, or arranged in a bar configuration make excellent display devices for the outputs of a system. A single LED can be used as an ON/OFF indicator and, if required, a flashing output can be set up by the software. For this a simple transistor interface (*fig. 7.9*) is more than adequate. Note the use of series resistors to limit both the current into the transistor switch and through the diode.

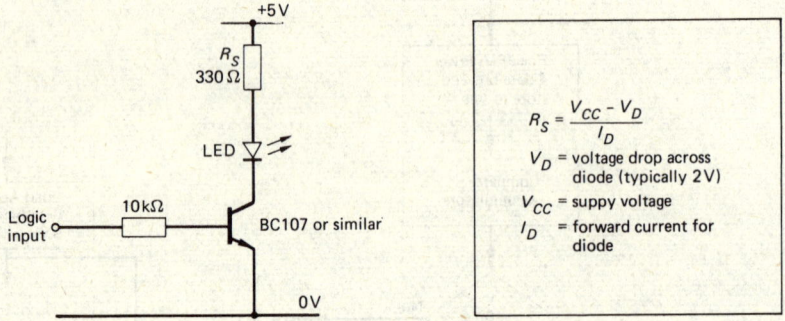

Fig. 7.9 Simple interface to drive an LED

$$R_S = \frac{V_{CC} - V_D}{I_D}$$

V_D = voltage drop across diode (typically 2V)
V_{CC} = suppy voltage
I_D = forward current for diode

When more than two or three LEDs are used, it may be better to use an IC to perform the buffer action between the port and the diodes. A TTL hex. inverting open collector buffer such as the 7406 makes a good choice (*fig. 7.10*). Again a series resistor must be used to limit the current through each diode. In the circuit, the LED comes on when a logic 1 is applied to the gate input. Take care when using this arrangement that the total chip current does not exceed 30 mA; this could happen if all six gates are used to drive LEDs and if more than three of the LEDs are switched on at the same time. To

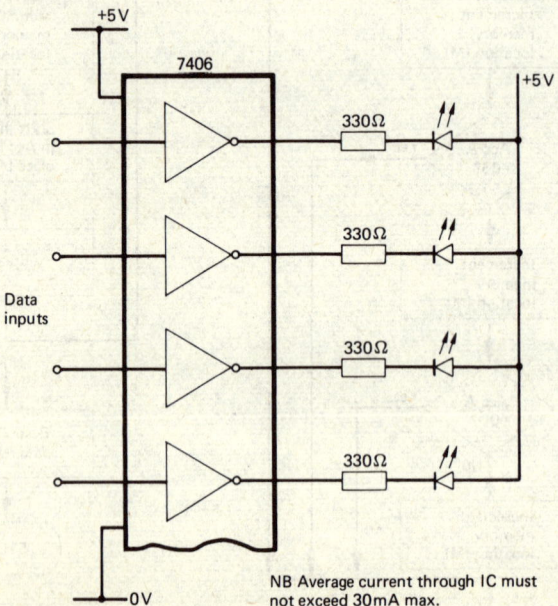

Fig. 7.10 TTL hex. inverting buffer

NB Average current through IC must not exceed 30 mA max.

Fig. 7.11 Driving a 7-segment common anode display

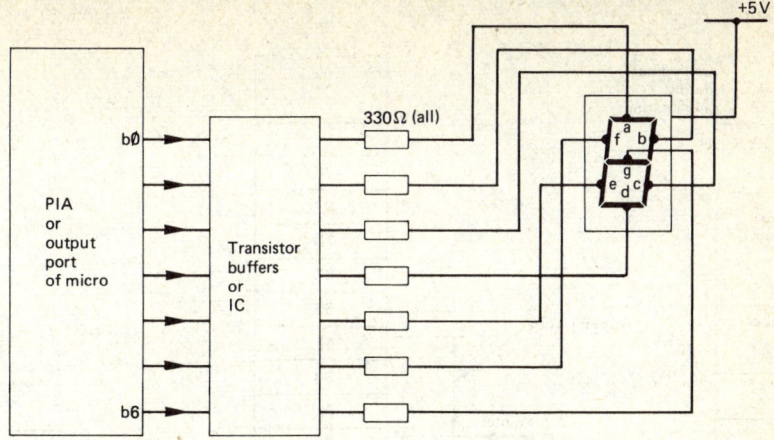

Display	g	f	e	d	c	b	a	Output Code
0	0	1	1	1	1	1	1	$3F
1	0	0	0	0	1	1	0	$06
2	1	0	1	1	0	1	1	$5B
3	1	0	0	1	1	1	1	$4F
4	1	1	0	0	1	1	0	$66
5	1	1	0	1	1	0	1	$6D
6	1	1	1	1	1	0	1	$7D
7	0	0	0	0	1	1	1	$07
8	1	1	1	1	1	1	1	$7F
9	1	1	0	0	1	1	1	$67
A	1	1	1	0	1	1	1	$77
b	1	1	1	1	1	0	0	$7C
C	0	1	1	1	0	0	1	$39
d	1	0	1	1	1	1	0	$5E
E	1	1	1	1	0	0	1	$79
F	1	1	1	0	0	0	1	$71

avoid this problem, use a multiplexed system where the diodes are switched on in sequence at a rate that is well above a value that would cause display flicker.

Seven-segment displays with either common anodes or common cathodes (*fig. 7.11*) can also be driven via transistor or IC buffers from an 8-bit port. Since each segment will require about 10 mA, the total current with all segments on is in the region of 80 mA, so care must be taken to ensure that the buffer IC is not overloaded. The segments are labelled a, b, c, d, e, f and g and it is possible to get all the hexadecimal numerals and letters displayed. The software, unless an external decoder IC is used, must do the job of producing the correct output code. For the common anode type, the buffer must receive a logic 1 to light a segment. Thus if the numeral 2 is required to be displayed then segments a, b, d, e and g must be on (logic 1 required) and segments c and f off. The output at the port must be $5B. As shown in the table there is a unique code for each hexadecimal character and these codes can be held in a look-up table in RAM or ROM ready to be outputted as

168 Practical Interface Circuits for Micros

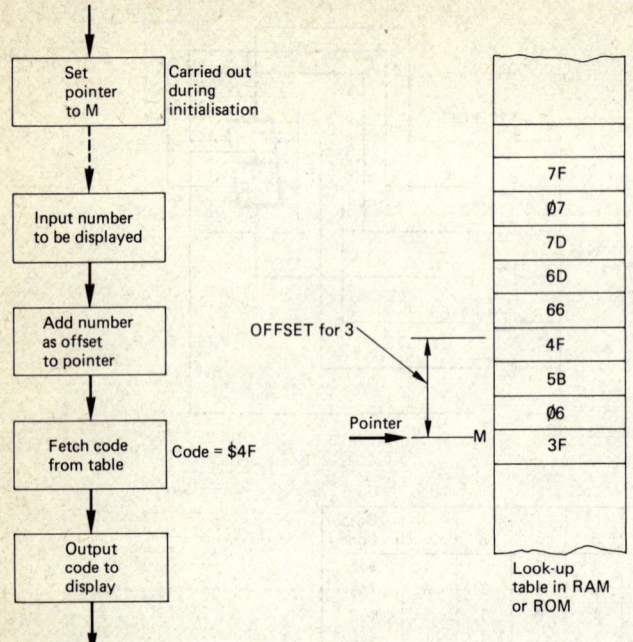

Fig. 7.12 Flowchart for outputting the code for a 7-segment display

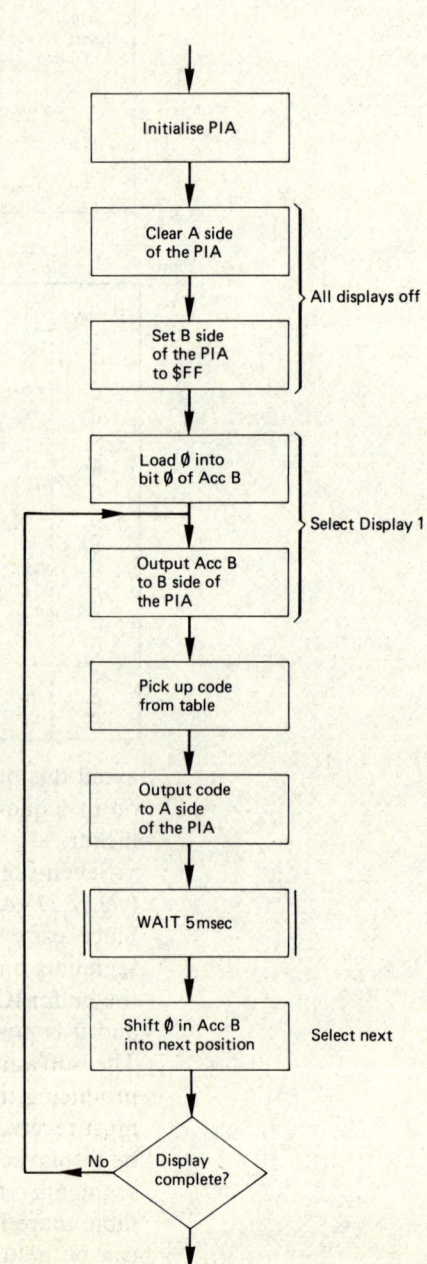

Fig. 7.14 Outline in flowchart form of the actions required in the multiplexed display

Application Examples 169

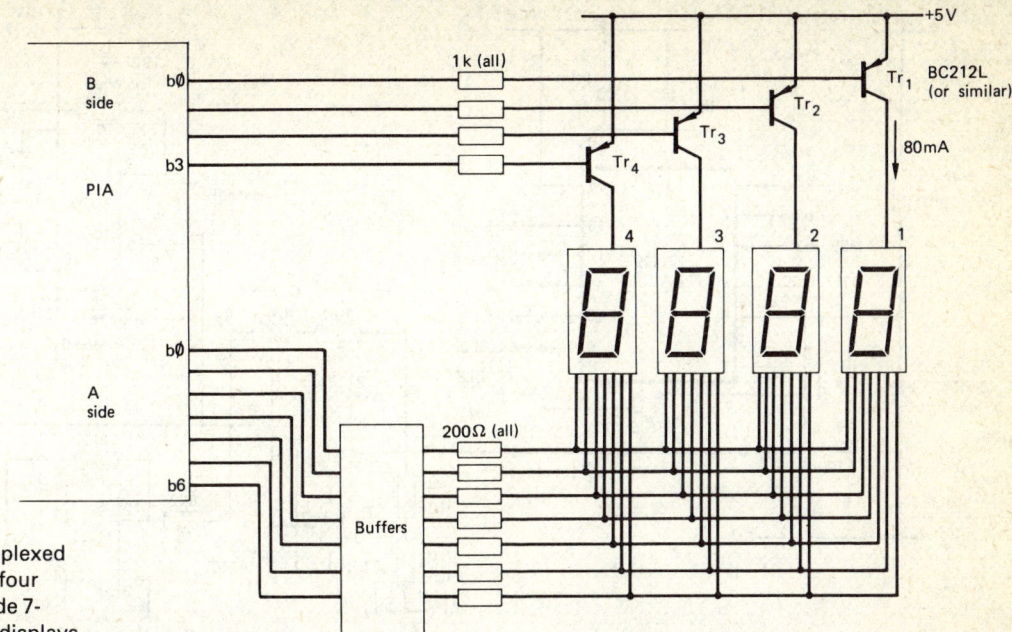

Fig. 7.13 Multiplexed display using four common anode 7-segment LED displays

required. The segment code table is accessed and the code is selected by using the number or letter required as an offset for the index register. This is illustrated as a flowchart in *fig. 7.12*.

When more than one seven-segment unit is used to display data, a multiplexed system is essential. The basic outline is shown in *fig. 7.13*. Only one display is energised at a time. This means that the same set of lines to the segments can be used for all the displays. For the common anode displays, transistors Tr_1, Tr_2, Tr_3 and Tr_4 will be switched on to supply the current required to the displays in a set sequence. The drive is shown to be from the B side of a 6821 PIA but any port can be used provided that it can sink about 5 mA. In the high state all four transistors are off and then one display is energised at a time by outputting a logic ∅ to the B output register and rotating this zero through the four bits. While a display is energised, the required numeral or letter is set up by outputting the appropriate code to the PIA data register A. This is buffered to switch the selected segments of the display that is on. The program or subroutine for display drive must continually refresh the display at a rate which will not cause flicker, say 50 times a second, and a time delay must be included to hold the output steady on each display to give sufficient light level. The sequence of actions is shown in flowchart form in *fig. 7.14*. The character codes can be held in a look-up table in the same way as for one display unit.

170 Practical Interface Circuits for Micros

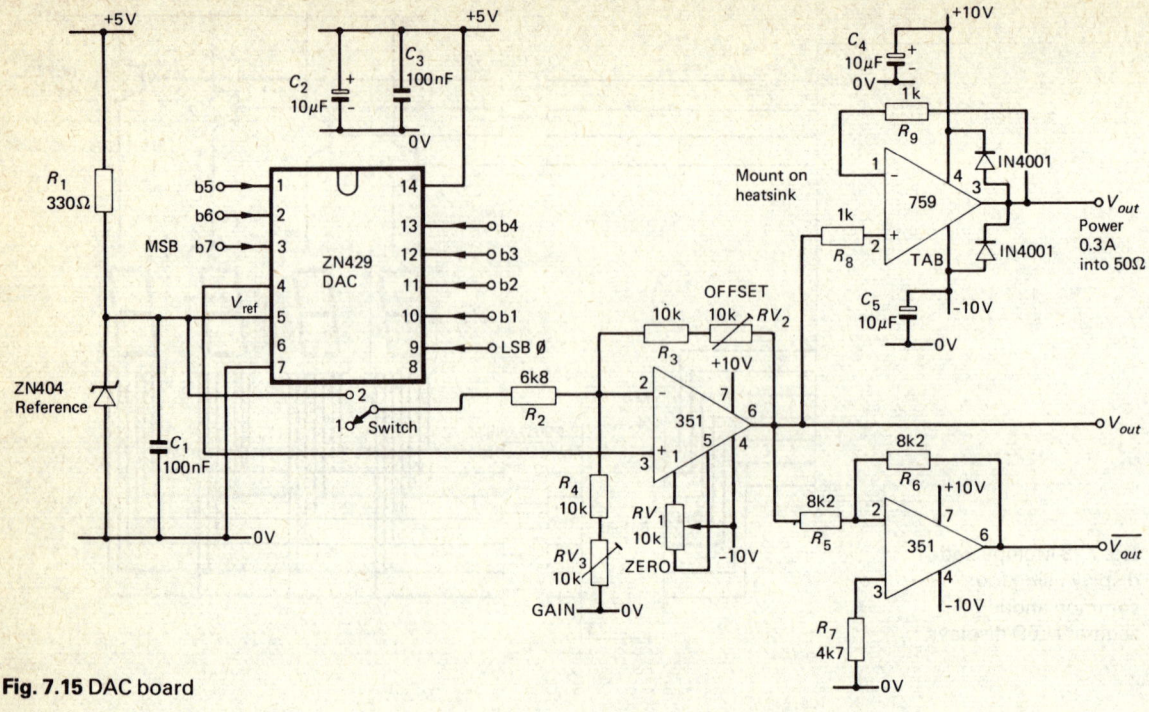

Fig. 7.15 DAC board

7.4 A General-purpose DAC Board

Some sort of standard interface board may be useful if a micro is frequently required to form the basis of many different systems. Instead of continually redesigning the convertors (DAC and ADC), general-purpose units can be employed with their use tailored to each new situation. Here a standard DAC board (fig. 7.15) is made up from a relatively inexpensive DAC chip, the Ferranti ZN429, together with a stable reference provided by the ZN404. The 8-bit ZN429 was previously discussed in Chapter 5, section 5.3.

It has an accuracy of 8 bits, runs from a single +5 V supply, and has a typical settling time of 1 μsec. The ZN404 provides a stable 2.45 V reference (2.38 V min to 2.52 V max) with the current through it set to approximately 8 mA by R_1. The analog output voltage from the ZN429 will therefore be 2.510 V with all bits set to logic 1. This analog output is buffered by two 351 op-amps to give V_{out} and $\overline{V_{out}}$. In addition, a power op-amp, the 759, provides a fairly high power output sufficient to drive small motors and other similar loads. The output can be set to give unipolar voltage, i.e. from zero to +5 V at V_{out}, or bipolar, i.e. from −5 V to +5 V, by setting the switch (or link) to positions 1 and 2 respectively. In order to eliminate any offset error and to set the gain, three presets are shown. These will be set up as follows (switch position 1):

1 Set all digital input to ∅ and adjust RV_1 until all output voltages are zero volts.
2 Set all digital inputs to 1 and adjust RV_3 (and RV_2) to give an output of $V_{FS} - 1\text{LSB}$, i.e. $+4.980\,\text{V}$.
3 Set the MSB of the digital input to 1 and all other bits to ∅. The output should be $2.50\,\text{V}$. No adjustments should be necessary, but if required repeat tests 1 and 2.

For bipolar operation set the switch to position 2, then:

1 Set digital inputs so the MSB = 1 and all other bits are zero. Adjust RV_1 to give zero volts.
2 Set digital inputs all to logic ∅. Adjust RV_2 to give $-5.0\,\text{V}$ output.
3 Set digital inputs all to logic 1. Adjust RV_3 to give $+4.98\,\text{V}$ output.

Following this, the operation of the interface can be tested by outputting an ascending digital weighted output with a short delay between, and checking that a sawtooth waveform is generated at the various outputs. A typical program segment for one ramp would be (using 6800):

```
          CLR A              Clear Acc A
STORE     STA A   $PORT      Output to DAC
          INC A              Increment Acc A
          JSR     DELAY      Go to sub delay
          CMP A   #$FF       Max output?
          BNE     STORE      Loop back
```

Either this could be used to produce a continuous waveform by branching back to the first statement or a counter could be used to output a fixed number of cycles.

7.5 A General-purpose ADC Board

The previous DAC circuit can be used together with a high-speed comparator and software to make a successive approximation ADC (see page 122), and this may be a good solution for converting analog inputs which are changing at relatively high rates. In many situations, the signals from transducers are varying in a slow fashion and a simple ramp and counter type ADC will be perfectly adequate. A conversion time of say 2 msec allows the micro to sample an analog input nearly 500 times a second. The circuit of *fig. 7.16a* is built around one of the most commonly available DAC/ADC chips, the ZN425E, and uses the ramp and counter principle. This principle is fully discussed in Chapter 1 (pages 15 to 21) but some further detail will be given here.

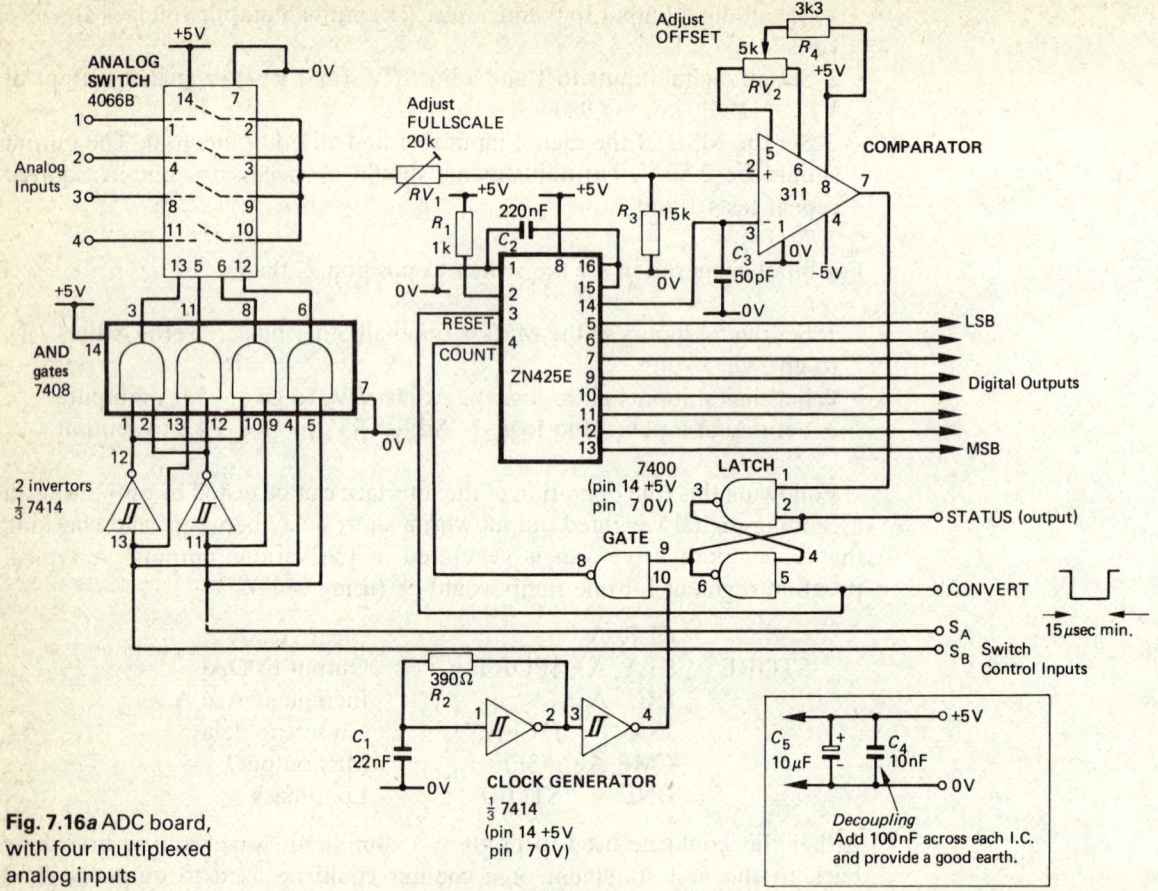

Fig. 7.16a ADC board, with four multiplexed analog inputs

The four analog inputs to the ADC chip are switched by a multiplexing circuit which uses some TTL gates and a CMOS 4066B quad analog switch IC. Switch control is by the pins labelled S_A and S_B which are decoded using two invertors from a 7414 and the four AND gates of a 7408. The sequence follows the table given below to ensure that only one switch is on at any one time. This allows up to four analog sensors to be used with the same ADC.

Inputs		Drive to operate Analog switches			
S_A	S_B	1	2	3	4
0	0	1	0	0	0
0	1	0	1	0	0
1	0	0	0	1	0
1	1	0	0	0	1

Here a logic 1 will operate the associated switch

The switches can be set up to select each analog input either by using software control, and this does give the arrangement maximum flexibility, or by using some hard-wired sequence generator synchronised to the status pulse

Fig. 7.16b Hardware circuit to operate the switch

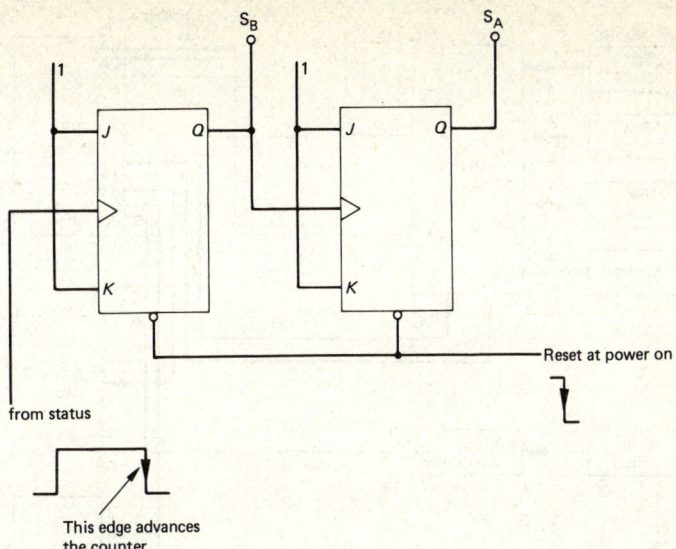

trailing edge. A circuit using two J-K bistables would be one way of doing this as shown in *fig. 7.16b*. Note that a power-on reset would be necessary to clear the bistables so that they start addressing with S_A and S_B both logic $\emptyset$ to switch input 1 first.

The selected analog input is applied via RV_1 to the non-inverting input of a fast comparator and is compared to the ramp generated by the ZN425E. You will recall that the ZN425E has an internal counter and an R-2R network that converts the state of the counter into a ramp output. Initially a negative going "convert" pulse (minimum width 15 μsec) sets the status latch and resets the internal counter of the ZN425E. Following this, clock pulses are gated to the count input of the chip and the internal circuits accumulate counts. These are converted internally by the R-2R ladder network to give a ramp output at pin 14. When the ramp just exceeds the level of the analog input, the comparator output switches low and resets the latch. The status returns to logic $\emptyset$ to indicate that the conversion has been completed and no more pulses from the clock are allowed through to the ZN425E. The digital value attained by the counter is then equivalent to the level of the analog input. The waveforms at various test points are illustrated on page 17. The conversion time will be a maximum when any analog input is near full-scale value (+5 V) and is set by the clock generator frequency according to the formula

$$\text{Max. conversion time} = \frac{256}{f_c} \text{ seconds}$$

With the clock set to 200 kHz, the maximum conversion time is then 1.28 msec. Allowing for a reset and read time totalling say 50 μsec, and considering the fact that there are four analog inputs, it is possible for each input to be sampled at a maximum rate of at least 180 times per second. The method for connecting the ADC to a micro together with suitable programming techniques is discussed in the system example in the next section.

174 Practical Interface Circuits for Micros

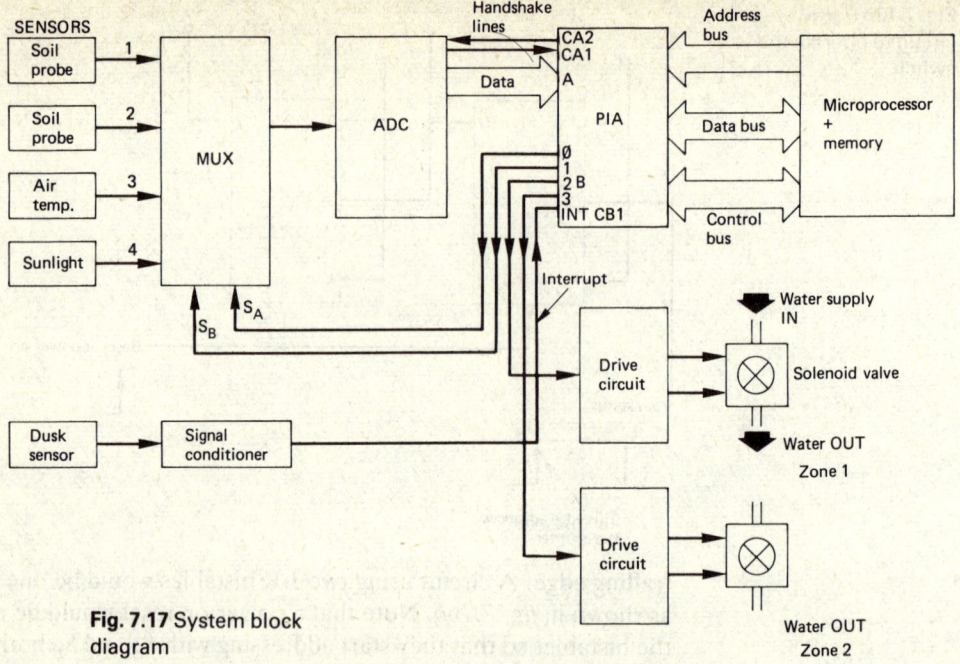

Fig. 7.17 System block diagram

7.6 System Example

What follows is the outline of a system, using a microcomputer, designed to automatically water a garden. The system is intended for use in an unattended garden when, for instance, the owner is away on holiday for a few weeks during the summer. Basically the required specification is as follows:

a) To collect data during the hours of daylight on soil moisture, air temperature and sunlight. For increased reliability two sensors are to be used for soil moisture.
b) Watering is to be carried out at dusk, with the amount of water supplied on a time basis computed from the stored data. The watering times are to be 10 min, 20 min, 30 min, 45 min and 60 min.

The block diagram of *fig. 7.17* shows that the four input sensors are switched in turn to the ADC under software control, with use being made of the general-purpose ADC board discussed in the previous section. The ADC is shown linked to the microcomputer via a PIA but the arrangement could be made to operate with any other type of parallel interface. The output signals from the micro are interfaced via isolating circuits to two heavy duty solenoid valves which control the water outlet to Zone 1 and Zone 2. The zoning allows additional watering for such things as hanging baskets and tubs. A light sensor, via a signal conditioning circuit, gives the pulse to indicate that it is dusk and the micro under program control computes the required time for the watering action. I do not propose to give the print outs of the program for this, but instead to simply outline in flowchart form the type of structure

Application Examples 175

required. The type of program can of course be tailored to particular situations. We shall pay more attention to such things as the choice of sensors and the way in which the signals from these devices are conditioned for the ADC. In addition, detail will be given of the design for the interface between the PIA and the solenoids.

The output signals from the various sensors do not have to be highly accurate but should have good repeatability. In any case the reliability of the sensors is one of the prime considerations. For the moisture sensors, a probe based on a standard jack-plug can be used, with the resistance of the soil sample being the measured quantity. The resistance will fall as the soil gets

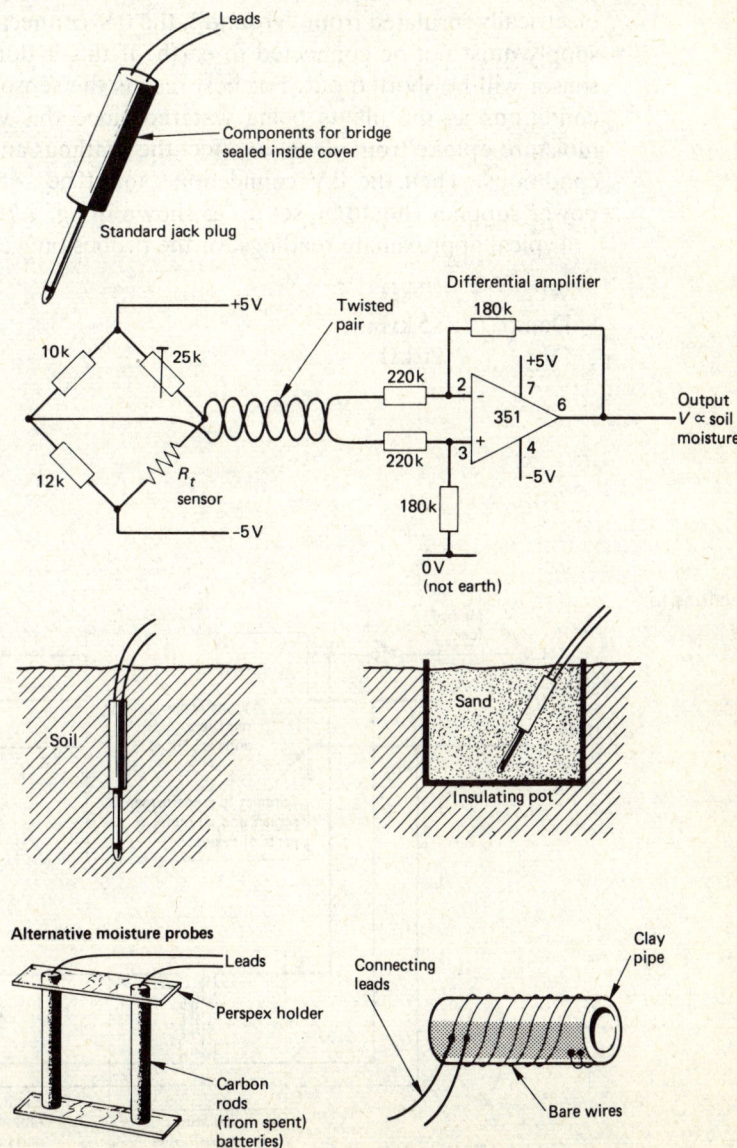

Fig. 7.18 Moisture probe and circuit

176 Practical Interface Circuits for Micros

damp, but it must be realised that the local soil conditions will determine the wet/dry resistance figures and direct on-site calibration would be essential. The probe (*fig. 7.18*) as part of a bridge network can be either fixed in the soil surface or buried as shown. The enclosed components and connecting leads must be completely sealed against moisture. Since the probe may be positioned some distance from the ADC board, care has to be taken to ensure that the resulting input signals are not affected by interference and are therefore reliable indicators of wet/dry conditions. The voltage output from the simple bridge network is taken via a twisted pair of wires to a differential amplifier circuit. Before we go on to discuss typical readings it should be pointed out that, unless the probe is placed in a tub or in a soil area electrically insulated from "ground", the 0 V connection of the system power supply must not be connected to earth. If this is done, the signals from the sensor will be shorted out. For best results the sensor should be in the same conditions as the plants being watered since this will allow drainage and moisture uptake from plants to affect the readings and give a truer picture of conditions. Then the 0 V connection cannot be "earthed" and the system power supplies should be set up as shown in *fig. 7.19*.

Typical approximate readings for the probes on well-drained light soil are:

Wet	2 kΩ
Damp	5 kΩ
Dry	20 kΩ

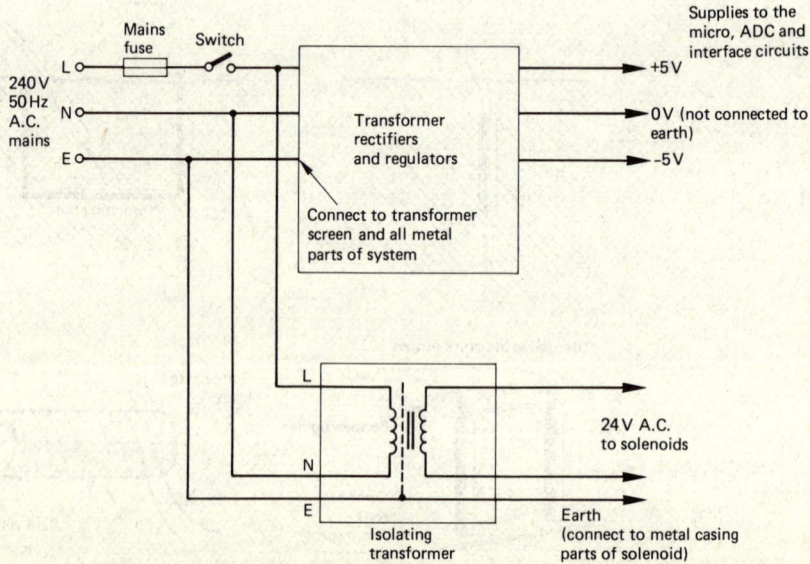

Fig. 7.19 Connections for power supplies

These resistance figures give an analog voltage from the differential amplifier in the range +3 V down to +100 mV (wet to dry).

Naturally there are many other ways of sensing soil moisture which may prove more reliable and accurate over a long time period than the probes based on a standard jack-plug. One of the main problems will be corrosion of the metal surface which will ultimately destroy the effectiveness of the probes. However a standard plug of this type is very cheap and the probes can therefore be replaced regularly. Another point is that improved reliability can be obtained either by using an a.c. supply to the probe bridge circuit, or by switching the d.c. supply only when the sensor is being read. These methods reduce corrosion but are not shown here since they further complicate the circuitry. The switching of the d.c. supply can be achieved to each of the four sensors in turn by using the same switching signal that operates the multiplex circuit.

The air temperature sensor (*fig. 7.20*) uses a forward biased silicon diode which is supplied with a constant current from Tr_1. As the temperature changes, the voltage across the diode will drop by about 2 mV/°C. The forward volt drop of 500 mV is eliminated by mixing the diode signal with a d.c. level of opposite polarity from the junction of R_4 and RV_1. The inverting op-amp circuit then gives an output signal of about 50 mV per °C change. For example, at an air temperature of +10°C, the amplifier output is say +3 V and this drops to +2.5 V at 20°C. This allows a range of temperatures from cool to fairly hot to be sensed. Again on-site calibration would be necessary.

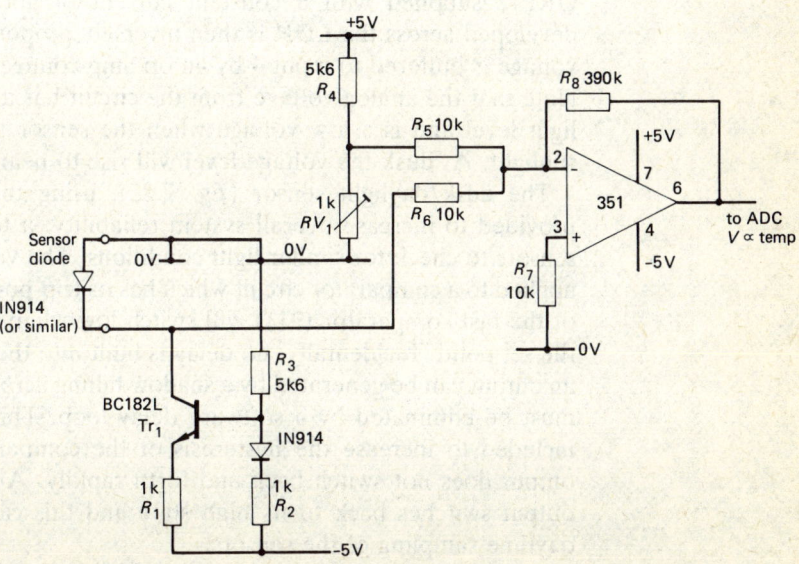

Fig. 7.20 Temperature sensing circuit

Fig. 7.21 Light-intensity detector

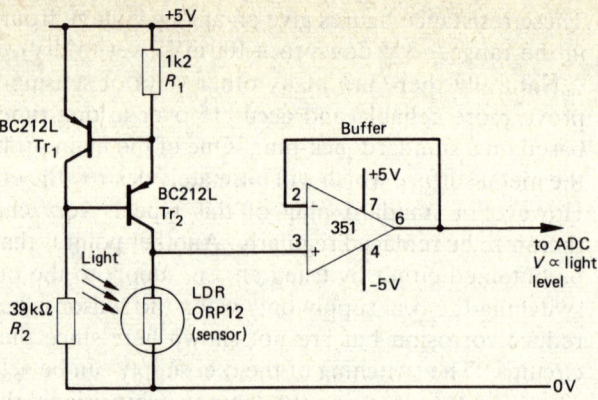

Fig. 7.22 Dusk/dawn sensor for interrupt

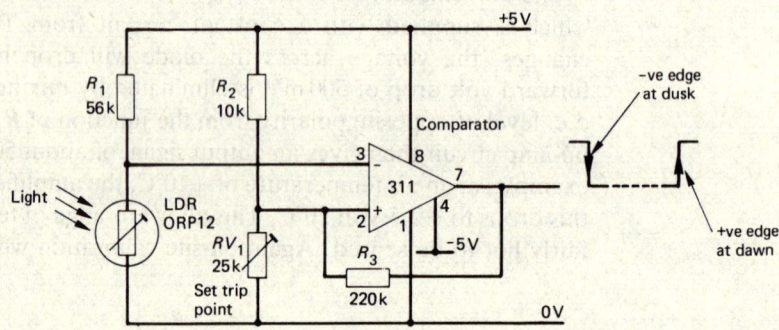

The sunlight detector (*fig. 7.21*) is a light-dependent resistor (LDR) type ORP12 supplied with a constant current of about 0.5 mA. The voltage developed across the LDR is then inversely proportional to light level. This voltage is buffered to input 4 by an op-amp connected as a voltage follower. Note that the analog voltage from the circuit has a nonlinear relationship to light level and is a low voltage when the sensor is exposed to very strong sunlight. At dusk the voltage level will rise to nearly +5 V.

The dusk/daylight sensor (*fig. 7.22*), using the same type of LDR is provided to increase overall system reliability; a test can be made on both sensors to check for similar light conditions. The voltage across the sensor is applied to a comparator circuit which has its trip-point set by RV_1. The output of the fast comparator (311) will switch low when the light level falls below the set point. Incidentally, no delay is built into the circuit which means that an output can be generated by a shadow falling across the sensor. Such effects must be eliminated by a software delay loop. This is discussed later. R_3 is included to increase the hysteresis of the comparator so that at dusk the output does not switch back and forth rapidly. At daylight the comparator output switches back to its high state and this can be used to initiate the daytime sampling of the sensors.

Fig. 7.23 Output interface circuit

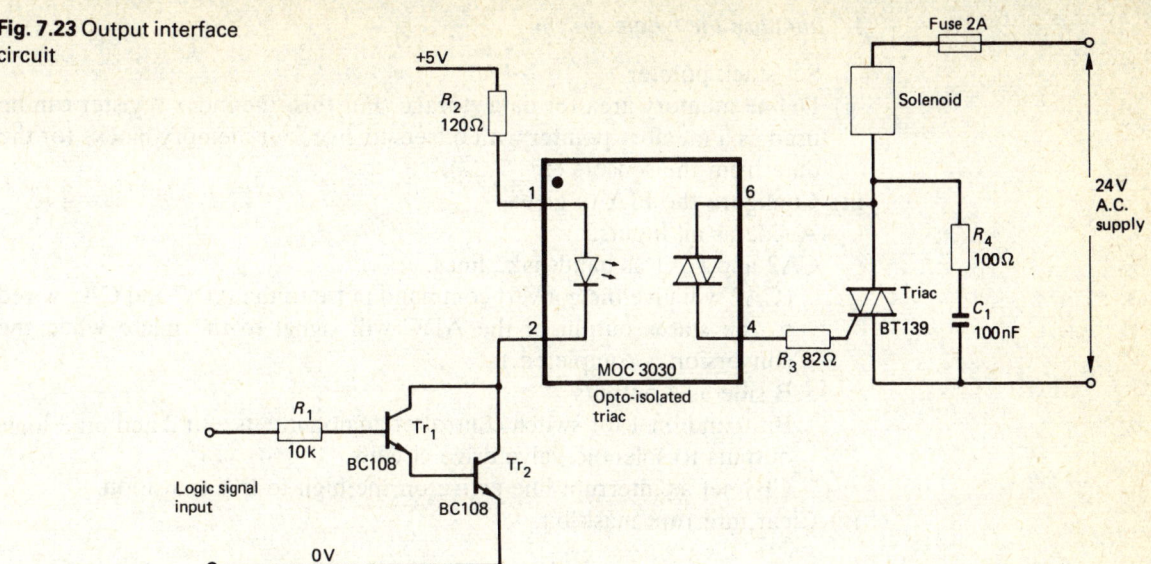

Before we go on to discuss the system operation and software requirements, we consider the design of the output interface (*fig. 7.23*). The solenoid valve is an a.c. type running from a 24 V r.m.s. a.c. supply derived from the mains via an isolating transformer. The logic signal from the micro output port can be either logic 1 or logic ∅. With the circuit shown, a logic 1 will switch the solenoid on. The Darlington transistors operate, causing the LED in the opto-isolator to conduct. This in turn fires the light-sensitive triac and the main triac, a BT139, conducts to pass current and operate the solenoid. The a.c. supply is completely isolated from the micro logic signals which is an important feature of any system such as this. A snubber circuit (R_4 and C_1) is fitted across the triac to prevent transient switching spikes from damaging the triac or causing it to switch when not required to do so.

Looking at the overall system and assuming that the program data can be input to the micro from a keyboard, or that the program is held in EPROM, we can now investigate the various tasks to be carried out. An algorithm for this is as follows (this assumes a 6800-based system).

1 *Initialise the whole system*

  i) Set stack pointer.
 ii) Define memory area for data storage. For this, the index register can be used as a memory pointer with offsets to give four memory blocks for the data from the sensors (*fig. 7.24*).
iii) Configure the PIA to give:
    A side as all inputs.
    CA2 and CA1 as handshake lines.
        (CA2 will give the convert command pulse to the ADC and CA1 wired to the status output of the ADC will signal to the micro when the conversion is completed.)
    B side as 4 outputs.
    Bit 0 and bit 1 for switch control of analog inputs. Bit 2 and bit 3 logic outputs to solenoid valve drive circuits.
    CB1 set as interrupt line active on the high to low transition.
 iv) Clear interrupt mask bit.

2 *During daylight*

  i) Read output of each sensor in turn and store the data in memory using the pointer.
 ii) Wait 30 minutes.
iii) Repeat read and storage of analog data.

3 *Following interrupt at dusk*

  i) Wait 5 minutes.
 ii) Recheck interrupt sensor; if still low, then
iii) Test light level output from light sensor 4 to confirm dusk condition, otherwise return from interrupt.

Assuming interrupt is valid proceed to **4**.

4 *Computation*

  i) Sort data on soil moisture (zone 1) and compute wet/dry level.
 ii) Repeat i) for wet/dry level in zone 2.
iii) Sort data on air temperatures and compute average value.
 iv) Sort data on sunlight and compute average level.
  v) Using the results of i), ii), iii) and iv) decide time for watering action for the two zones. Then clear data memory. Pass time value to output water action program.

5 *Watering*

  i) If no watering required move to iii), otherwise to ii).
 ii) Output water to zone 1 and zone 2 for time computed.
iii) Reconfigure PIA to allow for interrupt at CB2 on positive edge.
 iv) Wait for interrupt; following interrupt proceed to read sensors at **2** i).

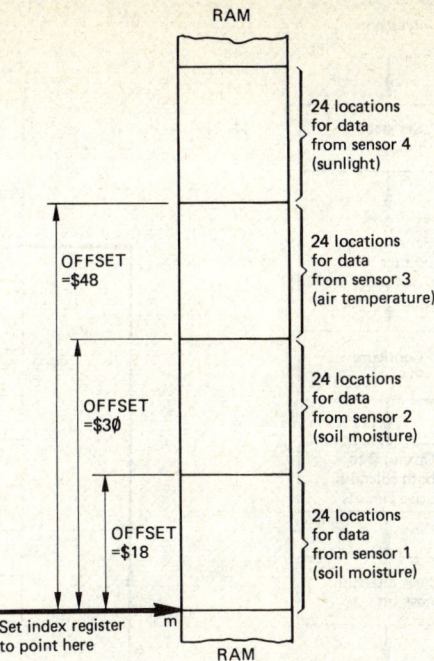

Fig. 7.24 Memory map for data storage

The program can be as simple or as complex as one wishes. For instance, after **5** ii) the state of both moisture sensors could be checked, after a suitable delay, to test that water has been output and some action could be taken if insufficient water had been supplied. This could occur if the water pressure falls for some reason.

The flowcharts for initialisation, data collection, data sorting, and output to solenoids are shown in *fig. 7.25a* to *e*. These are suggestions only. The time delays can be easily set up by software and an external timer is not necessary. For this application the micro can be considered to be dedicated to the task. Only minor modifications would be required to set up the system using any other type of microcomputer in place of the 6800.

182 Practical Interface Circuits for Micros

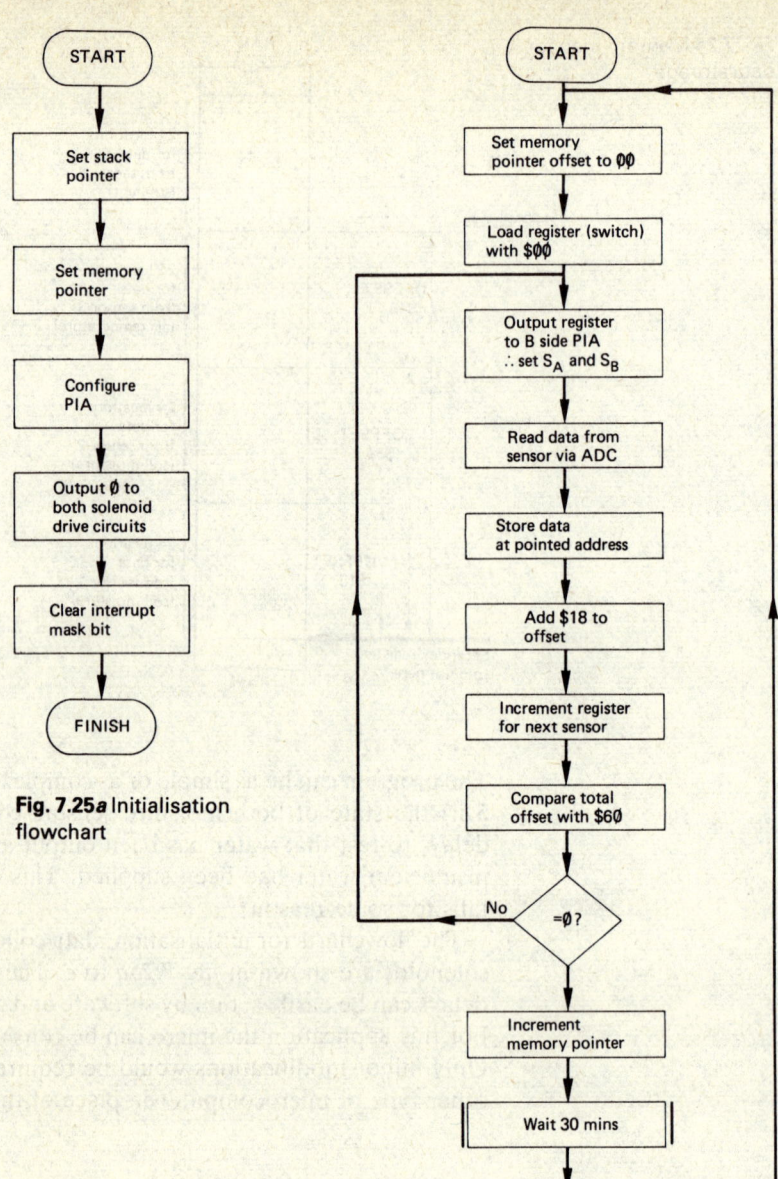

Fig. 7.25a Initialisation flowchart

Fig. 7.25b Data storage flowchart

Application Examples 183

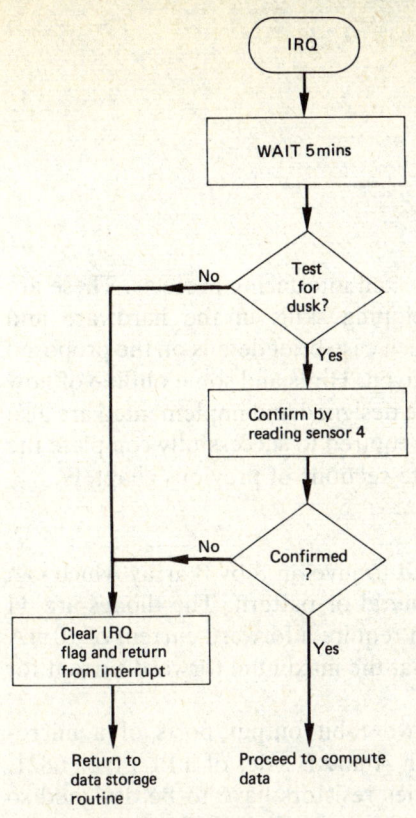

Fig. 7.25c Interrupt flowchart

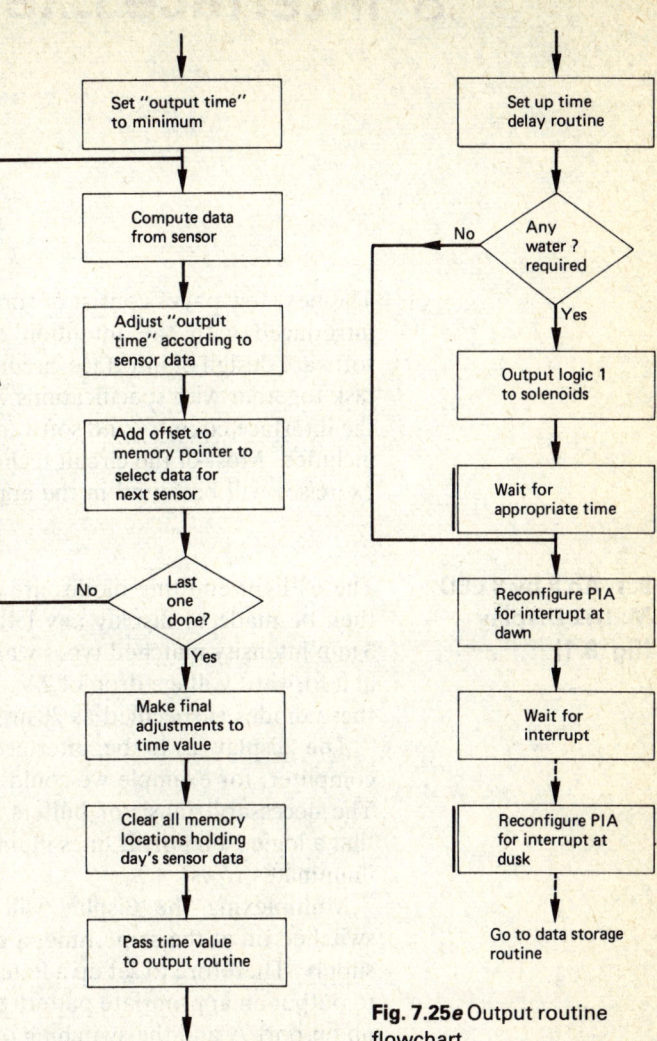

Fig 7.25d Data processing flowchart

Fig. 7.25e Output routine flowchart

8 Interface Exercises

The next few pages consist of some suggested interfacing projects. These are introduced with the intention of developing skills in the hardware and software design of interface circuits. In each case brief details on the proposed task together with specifications will be given. Hints and some outline of how the interface circuitry and software can be designed and implemented are also included. Most of the circuit techniques required to successfully complete the exercises will be found in the appropriate sections of previous chapters.

8.1 An 8 by 8 LED Matrix Display (fig. 8.1)

The 64 light-emitting diodes are arranged to give an 8 by 8 array which can then be made to display any letter, numeral or pattern. The diodes are TI 3 mm intensity matched types which each require a forward current of 10 mA at a forward voltage drop of 2 V. Note that the maximum forward current for these diodes is specified as 30 mA.

The display is to be interfaced to two 8-bit output ports of a microcomputer; for example we could use both A and B sides of a PIA type 6821. The necessary transistor buffers and series resistors have to be designed so that a logic 1 on port B lines illuminates columns and a logic ∅ on port A lines illuminates rows.

Multiplexing the display will be necessary because, if all diodes are switched on at the same time, a current of 640 mA will be required from the supply. Therefore to set up a letter or numeral, the software must be arranged to output an appropriate pattern to each column in turn; the pattern being set up on port A and the switching of the columns arranged via port B. The rate at which this is carried out must not produce any noticeable flicker.

Tasks
1 Design the interface buffers.
2 Write a program to test the circuits that will switch each column on in turn for say 0.5 sec and then clear the display.
3 Arrange for the software to
  a) Cause any one LED to flash at a rate of 2 per sec.
  b) Output the letter Z (or any other numeral or letter).

8.2 Waveform Generation Using the DAC Board (fig. 8.2)

The object of this exercise is to use the standard DAC interface board described in Chapter 7, together with a few external switches, to generate waveforms. The outputs are to be square, triangle and sine wave-shapes at spot frequencies of 400 Hz and 1 kHz. The frequencies and required

Fig. 8.1 An 8 by 8 LED matrix display

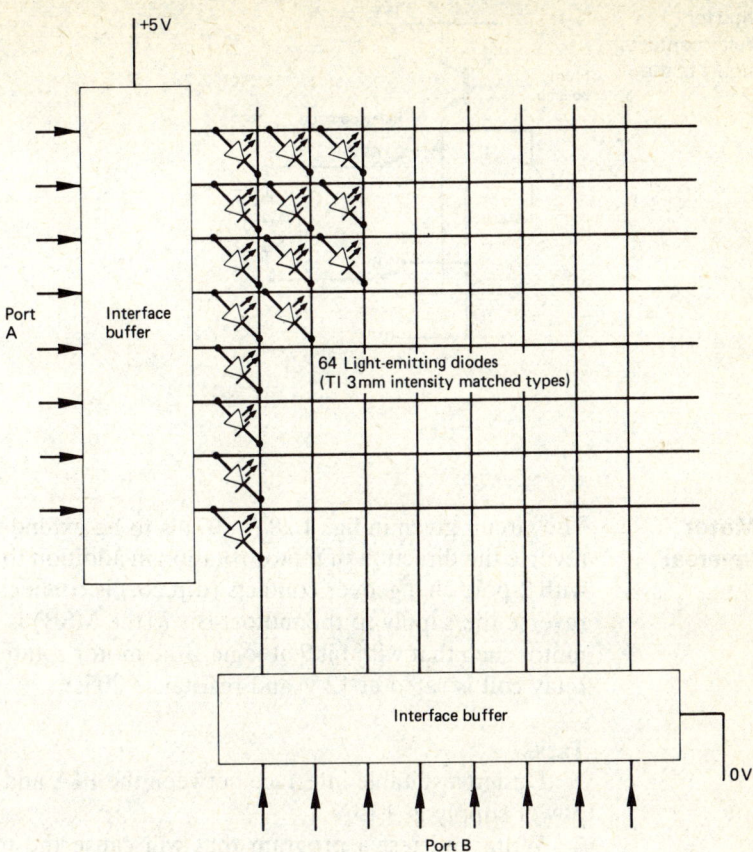

Fig. 8.2 Block diagram for waveform generation

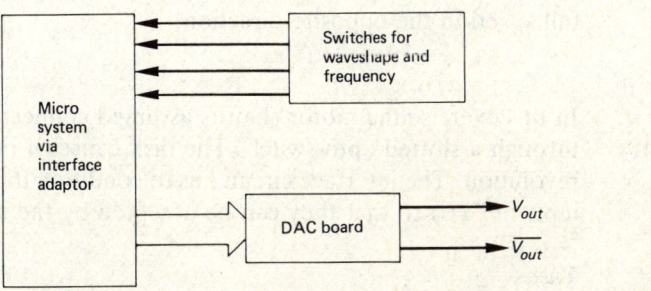

waveshapes are to be set up on switches using the A side of the PIA or the other 8-bit port of the microcomputer that you are using.

The switches will be required to be debounced and should be arranged to cause an interrupt when operated.

Time delay routines willl be necessary for generating the frequencies, and the sine wave can be formed by using a look-up table held in memory of values for one quarter cycle.

Can you extend the design so that two fixed levels of amplitude can also be set up?

Fig. 8.3 Method for connecting relay contacts to reverse supply to the motor

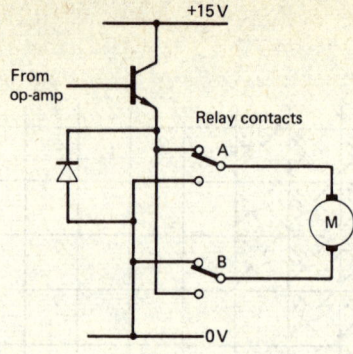

8.3 D.C. Motor Control Reversal

The circuit given in fig. 4.28 (p. 97) is to be extended so that it is possible to reverse the direction of motor rotation in addition to setting its speed. A relay with 2-pole changeover contacts (d.p.co.) is connected as shown in *fig. 8.3* to reverse the supply to the motor. Bit 7 (the MSB) is to be used to control the motor such that with bit 7 at logic 1 the motor rotates counter-clockwise. The relay coil is rated at 12 V and resistance 205 Ω.

Tasks
1 Design a suitable interface between the PIA and the relay. (The available power supply is +15 V.)
2 Write and test a program that will cause the motor to run from an off position to full speed in a clockwise direction, switch off, and then run up to full speed in the opposite direction.

8.4 D.C. Motor Control—Addition of Counter

In this exercise the motor shaft is assumed connected to a disk which moves through a slotted opto-switch. The disk causes 4 pulses to be generated per revolution. The interface circuit has to condition the pulses into suitable logic inputs (TTL) so that they can be accepted by the micro and counted.

Tasks
1 Design a suitable interface and conditioning circuit. A standard TTL Schmitt IC should be used to condition the pulse. The revolutions could be counted by either using additional hardware in the form of a counter IC or by interrupting the micro.
2 Write and test a program that enables a fixed number of revolutions (up to 255) to be output to the motor.

The basics of this system could be extended to provide closed loop control of the motor speed. In this case several more pulses per revolution would be required. This means that the disk will require say 64 printed dark lines.

8.5 Light Level Control

Four filament lamps rated at 12 V and 2.2 W are used to illuminate an enclosed space. The illumination has to be set and held constant at various levels during a test sequence, from a low value through to a maximum. A light sensor is used to detect the light level within the enclosure. The test sequence is not rapid, taking several minutes to complete a run.

The lamps are to be controlled using a pulse width modulated technique, and are switched on and off using power transistors or power FETs. The frequency of the switching waveform must not cause any flicker.

An additional sensor is used to monitor the temperature inside the enclosure, and if the temperature exceeds a pre-set value (75°C) then all lamps must be switched off and an alarm given.

Tasks
1 Draw a block diagram of the required system.
2 Specify the types of sensor to be used for light level and enclosure temperature measurement.
3 Design and test all the necessary interface circuits.
4 Write and test the software to give a sequence of 4 light levels each lasting 30 sec.
5 Design and test the alarm interrupt circuit.
6 A display using 2 LED segment units is to be included to give an output showing the illumination. Modify your circuit and software to incorporate this.

Index

Accumulator 132
Accuracy 74
ACIA 151
Address/decoding 146
Addressing modes (6800/6802) 133
Analog circuits 36, 45
Analog signal 12
Analog-to-digital convertor (ADC) 4, 16, 116
   general-purpose type 171
AND instruction 143
Assembly language 139

Block Diagram 2, 11
Bistable 65
   D type 66
   JK type 68
   RS type 65
Buffer 37
   analog 37
   tri-state 64

Clock pulse circuit 51, 129
Closed loop control 10
Comparator 38
Complementary MOS Logic (CMOS) 61
Condition code register 133
Counters 70

Darlington 42, 93
Debounce 158
Decoupling 60
Delay/subroutine 150
Digital circuits 52
Digital signals 12
Digital systems 13
Digital-to-analog convertor 105
   general-purpose board 170
Diode 29
Display driver 166

Errors (in conversion) 105

FET 33

Gates 53

Handshake 7, 17
Heat sink 47

Initialising 22, 147
Index register 132
Instruction set (6800/6802) 133
Interface 2
Interface adaptors 127
Interfacing between logic 63
Interrupts 6, 135, 138

Inverting amplifier 39

Linear variable differential transformer 88
Logic gates 53
Logic families 59
Look-up table 24

Masking 143
Microprocessor (6800/6802) 128
Motor 95
Multiplex 4, 167

Op-amp 36
Open loop 10
Opto-electronic devices 83
Opto-switch 86
Oscillator 45

Phase control 8
Photo-cell 84
Photo-diode 85
PIA 22, 140
Polling 6
Power FET 34
Program counter 132

R-2R ladder 109
Relay 92

Sample and hold 126
Schmitt trigger 47
Semiconductors 27
Serial data format 152
Sensors 74
Shift register 69
Solenoid 94
Stack 138
Stack pointer 133
Stepper motor 98
Systems 1
Switch decoding 162

Thermistor 77
Thermocouple 39, 79
Thyristor 44
Timer (555) 49
Transistor 31
Tri-state 64
Triac 44
TTL 59

Unused gate inputs 62

VFET 34

Zener diode 26, 30